William Carmichael McIntosh

The Resources of the Sea,

as shown in the scientific experiments to test the effects of trawling and of the closure of certain areas off the Scottish shores

William Carmichael McIntosh

The Resources of the Sea,
as shown in the scientific experiments to test the effects of trawling and of the closure of certain areas off the Scottish shores

ISBN/EAN: 9783337240318

Printed in Europe, USA, Canada, Australia, Japan

Blue Shark, the type of a group which often ruins man's nets and hooks, and defies his influence.

[*Frontispiece.*

THE RESOURCES OF THE SEA

AS SHOWN IN THE SCIENTIFIC EXPERIMENTS
TO TEST THE EFFECTS OF TRAWLING AND
OF THE CLOSURE OF CERTAIN AREAS
OFF THE SCOTTISH SHORES.

BY

W. C. McINTOSH, M.D., LL.D., F.R.S., ETC.,

PROFESSOR OF NATURAL HISTORY IN THE UNIVERSITY OF
ST ANDREWS,
DIRECTOR OF THE MUSEUM AND OF THE GATTY MARINE LABORATORY.

ILLUSTRATED.

LONDON:
C. J. CLAY AND SONS,
CAMBRIDGE UNIVERSITY PRESS WAREHOUSE,
AVE MARIA LANE.
1899

DEDICATED

TO THE MEMORY OF

THOMAS LAYCOCK, M.D., ETC.,

PROFESSOR OF THE PRACTICE OF PHYSIC IN THE UNIVERSITY OF EDINBURGH,
AND PHYSICIAN TO THE QUEEN IN SCOTLAND,
A FAR-SEEING AND ACCOMPLISHED PHYSICIAN AND PSYCHOLOGIST,

ALEXANDER MACDUFF, ESQ.,

OF BONHARD,
THE IDEAL OF A REFINED AND SAGACIOUS COUNTY ADMINISTRATOR,

AND THE MANY VALUED FRIENDS IN THE COUNTY
AND CITY OF PERTH, 1860—1883,

BY THE AUTHOR.

PREFACE.

THE author has felt it an urgent duty, since the work for the Royal Commission under Lord Dalhousie, to watch the various experiments to test the effects of trawling and of the closure, the more especially as these had been originally suggested by himself. Judgment was withheld until all the necessary facts were available, and a sufficient margin of time had supervened for the thorough investigation and testing of the areas. Now, after the lapse of twelve years, and of fifteen since the original work was begun; after a study of all that had been done at home independently of these experiments, and in other maritime countries; after a prolonged investigation into the development and life-histories of the food-fishes; and after the general statistics of the fisheries had time to settle into a reliable condition, it has been deemed prudent to make a critical survey of the work, and to form conclusions on this important question.

It was hoped that opportunities would have been afforded for repeating in 1898, on the same dates and, as far as possible, under the same circumstances, the experiments of 1884, but the authorities did not appear to see either the way or the importance of such an enterprise. The suggestion was made in no narrow spirit, indeed its acceptance would have entailed very serious responsibilities on the author.

The great aim of the survey has been to search for truth in this complex subject, to weigh carefully every fact bearing on its solution, and to bring an unbiassed mind to the task. To the scientific investigator a conclusion favourable to the continuance of restrictive measures would have been as welcome as the reverse, provided his facts warranted it. Misinterpretation or overstraining of the deductions in any case is equally distasteful, and would be equally shortlived.

It is interesting that similar conclusions to those formed by the author were reached by Professor Huxley and others from a totally different stand-point, a fact which does not detract from the strength of the position. The opportunities for practical inquiries into every branch of the subject have been—during the fifteen years' inquiry—greater than at any former period in this country.

While pondering over all the facts, and fully conscious of the responsibility entailed, one relief has ever been present, and that is—perfect faith in the marvellous ways of Nature in the ocean, ways which enable her to cope, in regard to the food-fishes, with all the wonderful advances in apparatus for capture, and with the steady increase of population.

It was intended to have printed every table prepared during the ten years, but their number (108) was formidable, and hence only such as were essential for the appreciation of the facts have been given. In the preparation of these the author has been aided by a series of valued young friends, amongst whom the late Rev. R. Gillespie, M.A., the Rev. E. Teviotdale, Mr W. E. Collinge, Mr Alex Thom, Mr Thomas Cargill, Mr Frank M. Milne, and Mr A. F. Munro deserve special mention. The genial companionship of these gentlemen and their unwearied efforts were sources of sincere satisfaction. The heaviest share of the work fell to Mr Frank M. Milne, M.A.

The author has also to acknowledge the aid received in regard to the woodcuts of fishes from Dr Murie, LL.D., and Messrs Cassell, Petter, Galpin and Co., and for photographs from A. Wallace Brown, a name—for fully fifteen years—both welcome and familiar to every worker at the Marine Laboratory. To his valued colleague in the University, Dr Masterman, he is indebted for the preparation of the Index.

In recent times great advances, inaugurated by Prof. Baird, have been made in connection with the fisheries by the United States Fish-Commission, and carried on by a distinguished band of workers, amongst whom the names of Alex. Agassiz,

Whitman, Earle, Jordan, Tarleton Bean, Ryder, Brown-Goode, and others are conspicuous. Canada, under the vigorous hand of Sir Charles Tupper, made a new departure in its fisheries by the appointment of the able and experienced Prof. Prince. The labours of the French under Professors Marion and Pouchet, Dr Sauvage, M. Guitel, Dr Canu, P. Gourret, G. Roché, M. Odin, and others have largely added to our knowledge; as likewise have those of Prof. Hensen, Prof. Möbius, Prof. Brandt, Dr Heincke, Dr Ehrenbaum and many others in Germany. Valuable work has been done in Denmark by Capt. Drechsel, A. V. Ljungman, and Dr Petersen, while the same may be said of Prof. Hoffmann and Dr Hoek in the Netherlands. In Norway the labours of Prof. G. O. Sars are everywhere known, followed recently by those of Dr Hjort, and in this country also the hatching of marine fishes under Capt. Dannevig has been very successful; while in Sweden the names of the Malms, father and son, and Lundberg are familiar. In Belgium Prof. van Beneden's talents were invaluable—as his son's are now—in the department, and in Italy those of Prof. Giglioli. Russia has also made great progress under Dr O. Grimm, Dr Borodine, Dr Kniepovitch, and others, and Spain under Lieut. R. Vela. Lastly Japan is making active efforts to master its fisheries under Prof. Kishinouye.

In Australia and the Cape of Good Hope able workers are busy with the problems of the fisheries, for example, Prof. Haswell, Mr Lindsay Thompson, and lately Dr Saville Kent in the former, and Dr J. D. Gilchrist in the latter.

In our own country the earliest work was done at St Andrews, which has never lost touch with the subject; and the workers from which, *e.g.* Prof. E. Prince, W. L. Calderwood, E. W. L. Holt, Dr Scharff, Dr H. Wilson, Dr A. T. Masterman, W. E. Collinge, G. Sandeman, H. C. Williamson, J. R. Tosh, H. M. Kyle, and others have extended our knowledge of the subject in a noteworthy manner. The names of the late James Duncan Matthews and the late George Brook, on the staff of the Fishery Board for Scotland, again, will long be remembered for the excellence of their work in connection with the food-fishes. The Marine Biological Association at Plymouth has also made important advances by means of the able researches of W. Heape, J. T. Cunningham, G. C. Bourne, W. Bateson, W. Garstang, E. J. Allen and others. Lastly, under the Rev. W. S. Green and Prof. Haddon, an important survey, especially of the food-fishes in the deeper water, off the west coast of Ireland, was made by the St Andrews workers, Prof. E. Prince and E. W. L. Holt.

Prof. Herdman on the Lancashire coast has in recent years inaugurated modern methods in the fisheries with success, while Mr Meek at Cullercoats on the opposite shore has also made a commencement in the department.

In the criticism of the work of the Fishery Board for Scotland and its methods in regard to closure, it must not be supposed that the author has other feelings than those of respect for that body, from whom in former days he received much courtesy. It is sufficient to mention the names of Sir Thomas Boyd and Mr P. Esslemont—the former distinguished by his wide experience, the latter by his remarkable administrative powers, to indicate how ably the duty was done by each in the office of Chairman, with the support of such well-known members as Mr Maxtone Graham of Cultoquhey, Mr S. Williamson, Mr J. J. Grieve, Mr William Boyd, and Sir James Gibson Maitland of Howieton. The position and talents of the sheriffs, again, and their intimate acquaintance with fishery laws and administration, gave the Board both strength and prestige. The loss of the counsels of Sheriff Guthrie Smith, Sheriff Forbes Irvine, Sheriff Thoms and Sheriff Makechnie to the fisheries and the public cannot be over-estimated. To the energy and ability of the Board's Superintendent of Scientific Investigations (Dr T. W. Fulton) the author would also pay a just tribute of commendation.

Nothing, however, has been allowed to interfere with the faithful discharge, to the best of the author's ability, of a public duty, and in relation to a question of such vital importance in the department.

It is a source of regret that the popular statesman who was at the head of the Royal Commission (1883—85) has not lived to see the result of these experiments in which, from their commencement in the beginning of January 1884, he took so deep an interest. Even to the last he kept himself acquainted with their progress, and though, when he died in 1887, the systematic work of the "Garland" had just begun, he saw how necessary it was to test thoroughly and check previous observations. Indeed, when in 1884 any decided view was expressed, he was wont to add the caution "Remember others will follow, criticise and check every step taken." Yet, perhaps, his acutely sensitive mind, not to allude to his well-known sympathy with the liners, whose hardy, daring ways aroused his respect, and to whose appeals he ever cordially responded, might have felt a shade of disappointment at a result so different from the oft repeated views and wide-spread opinions of the fishing-community and the public.

GATTY MARINE LABORATORY,
ST ANDREWS.
15 *November*, 1898.

CONTENTS.

CHAPTER I.

CHAPTER II.

CHAPTER III.

CHAPTER IV.

CHAPTER V.

TABLES.

ST ANDREWS BAY.

FRITH OF FORTH.

MORAY FRITH.

FRITH OF CLYDE.

The Gatty Marine Laboratory, St Andrews.

To face p. xiv]

CHAPTER I.

INTRODUCTORY.

General Review of the Resources of the Sea and the Influence of Man thereon.

In this chapter is a general statement of facts which point to the conclusion that, with some exceptions, the fauna of the open sea, from its nature and environment, would appear, to a large extent, to be independent of man's influence.

In examining the resources of the sea[1] we are confronted with very different problems from those which meet us in the consideration of the fresh waters and the land. Moreover, this peculiar divergence has a tendency to mislead those whose experiences have been accumulated in surroundings in which all the operations of nature and of man are readily observed and easily understood.

It is but recently since we relied, for instance, on the knowledge derived from the effects of the non-protection of land-animals as a guide in dealing with the sea-fishes. It satisfied some to urge that man's influence had swept off the larger mammals from great areas, and at the present moment is rapidly diminishing the numbers of such forms as the right and other whales, the elephants, giraffes, elks, buffalos and bisons. Comparatively recent legislation in our own country has year by

[1] By the sea is meant the open ocean and the exposed shores, for enclosed seas, like the Mediterranean, are placed under different conditions as regards the fisheries. Consequently the effect of man's interference in certain cases is distinguishable—were it only in the single feature of the size of marine fishes offered for consumption.

year rendered the hare rarer than it has ever been previously, indeed it has swept this graceful denizen of the fields, this delight of every lover of nature, as well as unrivalled nourishment for the sick, almost out of certain counties, and may lead to its almost total extinction; the wolf has disappeared, while the badger is scarce, and where, we may ask, would the red deer, the roe and the fox have been if a measure of protection had not been meted out to them? Man's interference caused the extinction within a generation of the huge *Rhytina* (sea-cow), though it was more or less marine in habit, and in the present day the dugong and the manatee are diminishing and may follow their relative. As a rule, it is not difficult for man to influence land-animals whether for abundance or scarcity. Yet even here, under favourable surroundings, certain forms, as the rabbit in Australia, the lemmings in Norway and Sweden, the rats and mice of our houses, prove more than a match for his most potent checks. In the vegetable world, again, it is an arduous and often fruitless task to extirpate many plants—even with man's ingenuity and opportunities. Nature provides in the majority a vast number of seeds, far beyond what is required for the preservation of the species, indeed the older naturalists pointed to this as a wise provision for the support of animals. It is only when we come to the larger slow-growing kinds that the action of man is so destructive, and yet the number of seeds in the larger trees, as the pines, acacias and others, is very great, 30,000 for instance occurring in such as *Mimosa Lebbec.* Man's easy access and abundant opportunities are, however, very different from the conditions existing in the ocean.

Even in the domain of the air, which in its vastness excels that of the sea, the effects of man's interference—secondary though it may be—is often noteworthy, mainly because its denizens deposit their eggs or young on the surface of the land, and thus, though the adults may wing their way into the ether, and even feed therein, as in the case of the bats and swallows, the helpless young and eggs are within his reach. In the flightless birds the same rules apply as in the land-mammals, and how few of these now live! The moa, the great auk, and the dodo have gone, and the ostrich, emu, cassowary and kiwi

(*Apteryx*) will require protection to retain them. In a few instances, as in the penguins, remote haunts, active aquatic habits and great boldness have aided the species in warding off destruction and placed them in a different category. The agency of man quickly diminishes many of the larger birds of flight, and this notwithstanding legislative protection. Without the latter where would our wild swans, geese and ducks have been? where the capercaillie, black-cock, grouse, pheasant and partridge? The same lesson is learned when we contrast the former condition of the rook with the present one of modified protection. A few birds, as the sparrow, cushat and passenger-pigeon, appear to be exceptions under certain conditions; but these in the former case are due to the complications of civilized life itself, and in the latter often depend on immigration, and thus are seldom altogether beyond man's influence.

The air, however, differs from the land not only in its vastness, but in the fact that certain of its inhabitants, *e.g.* the insects, seem to defy man's power and ingenuity in limiting their numbers. Thus, is the house-fly less abundant now than in the days of the ancient Egyptians? Do the hordes of locusts, gnats, Hessian flies, mosquitoes, or other forms show marked diminution? Yet in all these instances the eggs are deposited on or near the surface of the earth and within man's reach. Even where elaborate measures are taken to encompass the destruction of a noxious form, *e.g.* the vine-parasite, *Phylloxera*, on plants under man's immediate care, how difficult is it, with all the resources of modern chemistry, to attain success. The same applies to the larger insects, such as wasps, the nests of which are easily reached.

From age to age these denizens of the air have kept their ground against all the forces man could bring against them, and yet they would have required aërial eggs to have placed them on a similar footing to most of the food-fishes of our waters.

The elaborate laws framed for the protection of the more valuable fishes of the fresh waters of our country are sufficient proofs of the care which is necessary for their protection. Diminution of these is readily caused by the various nets and other instruments used by man for capture. But just as—even in large areas of fresh water—the stock of fishes may be reduced to small dimensions by over-fishing, so restoration by artificial measures can be effectively carried out. In both respects, therefore, fresh waters offer a contrast to the sea.

When we come to the ocean the problems connected with man's influence on certain of its denizens assume a much greater degree of complexity. In the first place, though the almost boundless space of the air far surpasses the sea, the latter greatly exceeds the land, since it occupies about three-fourths of the earth's surface. Yet this gives but an imperfect comparison, for, whereas it is chiefly on the surface of the earth that terrestrial animals are found, it is altogether different with the ocean, the inhabitants of which not only people the bottom, but glide over the latter, frequent mid-water and the surface, indeed, may be said to be scattered everywhere throughout its mass as well as fringe its margins. Moreover, while we can pursue the mammal on land, entrap the fish in the stream or lake, or follow the flight of the bird to a certain extent in the air, it is otherwise with the sea. Its ever-changing and often opaque and tempestuous waters offer a barrier to the pursuit of its larger forms, even were it possible to track them to its distant abysses, while the more minute for the most part escape observation. Thus with all the skill and perseverance of ages much yet remains to be accomplished in regard to our knowledge of the sea and its inhabitants, both plant and animal. One feature, however, sufficiently distinguishes the sea from the air, viz. its being almost everywhere permeated by life—both plant and animal, and thus it affords a never-failing supply of food to its inhabitants, each of which finds in the surrounding water the nourishment best suited for it.

The resources of the ocean, however, are limited in the case of the large air-breathing forms pursued by man, such as the right whale or "bow-head," which has steadily decreased in numbers during the present century. The reckless slaughter of the young whales accompanying their dams, a sure method of capturing the unfortunate and solicitous mothers, has intensified the effect of this eager chase by various nations for whalebone and oil. Producing but a single young one at a birth, this huge and harmless mammal will probably disappear unless measures are taken for its preservation. The same may be said of the other whalebone whales which are pursued for profit, and of the dugong and manatee, the oil, skin and skeletons of which are of value. The huge Pacific grey whale (*Rhachinectes glaucus*) of the lagoons of the Californian coast has, indeed, been entirely destroyed by man.

The effect of the slaughter of hundreds of the ca'ing whale (*Globiocephalus melas*) is not so clear, but the xiphoid whales captured in the north seas for their oil are in greater danger. In no species has the inability of recuperation from constant attacks been better illustrated than in the sperm-whale, the numbers of which have been seriously diminished within recent times.

The effect of man's keen pursuit is also well shown in the seals, which are now so scarce on many parts of our own shores. Only in the more remote regions (as in the sound of Harris) are they to be met with in considerable numbers. Yet here and there the absence of interference and the presence of a favourite food may occasionally lead to their appearance in larger numbers, as at the mouth of the Tay. Man's effective influence on them in their headquarters in the Arctic seas is unchallenged, since their breeding places are on land.

These large air-breathing marine forms, therefore, occupy a special position, and make a contrast with the majority of the types we have now to consider.

Of great importance in connection with the subsequent groups is the plenitude and variety of marine plants. Familiar forms of these are the green, red and olive weeds of the rocky shores and tidal pools, and the grass-wrack of the muddy flats. In certain places indeed the perennial growth of the olive weeds was formerly of at least as much importance to the inhabitants as their grain-crops. The immense Sargasso-sea off the Azores, and the smaller ones in the Atlantic, again, have long been a source of interest and wonder. Various naturalists further have noted the discoloration of the sea from the abundance of microscopic plants, have observed the occurrence of vast masses of diatoms in the south Pacific, and as an ancient marine deposit on land; while others have secured many in the stomachs of ascidians.

Though it is not so long since the intimate connection between this ubiquitous plant-life as food for the higher marine forms has been prominently demonstrated, yet it had not altogether been overlooked. The herbivorous cetaceans, as the manatees and dugongs were called, are well-known instances. From time immemorial dulce, Irish moss, many other algæ, and edible swallows' nests (an algoid substance), have been esteemed by man; and certain land-animals, as pigs, cattle and ponies, eat with avidity tangles or other olive sea-weeds. Even the microscopic plants of the ocean (viz. diatoms) in their innutritious condition as a deposit on land (mountain-flour) were formerly mixed with the pounded inner bark of the Scotch fir to satisfy the simple appetites of the Norse peasantry.

The presence of the minuter forms of plant-life must have been familiar to fishermen and voyagers from early times, either as a coating on their nets, as causing discoloration of the sea, or stranded as a scum on the beach. The occurrence of diatoms and other minute forms has also been noted by scientific observers such as Sir Joseph Banks, Sir Joseph Hooker, Dr Gwyn Jeffreys and the various explorers of modern times, *e.g.* those on board the 'Challenger,' as well as the staff of the German Plankton Expedition. The great abundance of diatoms in the Arctic and still more in the Antarctic seas is well known, while in the depths of the Black Sea all else is stated to be absent but bacteria.

Detailed examination of the pelagic life of St Andrews Bay and the offshore waters of the east coast for many years showed the vast abundance and variety of minute plants at all seasons, from January to December[1]. Their occurrence in the stomachs of many invertebrates and even in some fishes further indicated the important part played by them. The presence of such as *Rhizosolenia* not only coloured the sea, but coated nets with an odoriferous layer. The intimate connection between the two great kingdoms was clearly pointed out as follows:—'It is a remarkable fact that it is primarily to plants in inshore

[1] *7th Ann. Report, S. F. B.*, 1889, pp. 259—310, 4 Plates.

waters that the abundance and variety of animals are in many respects due, especially if estuaries also debouch in the neighbourhood. Thus nowhere are the swarms of Sagittæ, Appendicularians, Crustaceans, and other groups of fish-food more conspicuous than in the midst of a sea teeming with diatoms, Rhizosoleniæ and other algoid structures.'......'Now this plant-life is specially rich in April and May, just when the larval and very young post-larval fishes appear more abundantly in the inshore waters, so that the cycle is nearly complete, viz., from the inorganic medium—through microscopical plant and larval crustacean—to the post-larval fish[1].' Similar views have been broached by Prof. Hensen of the German Expeditions and Dr George Murray of the British Museum; while Prof. Cleve, of Upsala, is of opinion that even the origin of the coastal currents in certain cases may be traced by their diatoms[2].

Besides, both adult and adolescent marine fishes such as sand-eels devour the green and greyish-green algæ of the Eden, even to the distention of their stomachs in May. Algæ, again, are not uncommon accompaniments of food in other fishes.

Broadly speaking, therefore, the sea has within its area a vast and ever-present source of nourishment for its teeming animals—a source altogether independent of the increment from fixed marine vegetation on shore or swept into it by rivers. This supply can in no way be affected by the action of man, who is as powerless to modify it as to modify the tides. So long as it remains, one of the most important factors for the safety of the food-fishes is secured.

In the ocean are immense numbers of Foraminifera and Radiolarians upon which the lower invertebrates prey both at the surface and the bottom. They in turn feed on diatoms and other simple plants and animals, and thus aid in completing the cycle from plant to fish. The records of the rocks show that many of both groups were as prevalent in the ancient seas

[1] Lecture, *Royal Instit. of Great Britain*, Friday, February 1, 1889, p. 10.

[2] Dr Petersen, of the Danish Zoological Station, has just issued an interesting addition to our knowledge of this subject—from observations in the Limfjord.

as in the modern. In the warmer areas, again, the water teems with the minute phosphorescent *Noctiluca*.

The Cilioflagellates, such as *Peridinium* and *Ceratium*, occur in great abundance throughout the year. In the warmer months these and the pelagic Infusoria (*e.g. Tintinnus*) are especially conspicuous at the surface. Their multiplication and that of other free and parasitic Infusorians—upon which many simple forms and even fishes feed—is beyond the sphere of human influence.

Amongst the lower invertebrates we meet with a group which in itself has been the source of a special fishery, viz. the sponges. From time immemorial the Levant and for a less period the Bahamas and Florida have furnished large supplies. Moreover, comparatively little care has been taken to propagate them by artificial means, and yet so nicely do all the surroundings fit them to continue the race that, after all these years, no sign of extinction is apparent. The fragments left on the rocks or stones as well as the pelagic larvæ which represent them in the water around have sufficed to fill up the gaps caused by the sponge-fishermen.

The non-commercial kinds of sponge, on the other hand, are free from special pursuit, being only captured by trawls and hooks to be again returned to the sea. The lacerated fragments or their contained eggs and larvæ enable each species to keep up its numbers in all our seas. Storms, moreover, toss many on the beach, but neither the one loss nor the other affects their abundance. Nature persistently carries out her measures of economy from age to age, and no more conspicuous example of this exists than the wide-spread disintegration of limestone rocks and the densest shells by the ceaseless borings of a sponge (*Cliona*).

In no group are the resources of the sea in regard to recuperation more prominently exhibited than in the hydroid zoophytes, jelly-fishes, anemones and corals. The hardihood of the common freshwater Hydra, so graphically told by Abraham Trembley, the old naturalist of Geneva, is characteristic of the majority of the race. The immense profusion of such zoophytes as *Obelia* in inshore waters and the rapidity of their growth show the ever ready resources of nature in restoring losses and spreading useful forms on every suitable site. Though it were possible in a given area to remove every vestige of such a form from the sea-bottom, a single summer tide would carry a sufficiency of little ripe jelly-fishes (*Hydromedusæ*) to repeople it. The countless swarms of these graceful and beautiful glassy creatures, which stretch for miles round our shores and far beyond into the open ocean, is one of the most striking features in marine life. From the eggs of these swimming jelly-fishes larvæ arise which by-and-by settle on rocks, stones, salmon stake-nets and other zoophytes, indeed, upon everything that affords a suitable hold, and rapidly grow into the plant-like original with which the cycle commenced. In their pelagic condition they are for the most part unaffected by any mode of fishing, though they are often beached in multitudes by the tide. Nature herself in another instance ordains an annual check to growth on the mussels of the Eden, for the dense and graceful tufts—heavily laden with young mussels—are swept off in October. But before this happens the spore-sacs of the zoophyte (*Gonothyrea*) have given rise to multitudes of pelagic young which ere long renew the feathery coating on the mussels.

Some of this group, again, such as *Corymorpha*, live immersed in sand, and are difficult to dislodge, even when specially sought after by trawl, dredge or net. Thus a large species obtained in considerable numbers on the ground-rope and trawl of the 'Medusa' in 1884 in St Andrews Bay has never been procured since, notwithstanding various efforts. Yet the free-swimming jelly-fishes it apparently throws off are common at certain seasons, showing that the parent-stock still remains. A rare plant (*e.g. Asplenium septentrionale* of Stenton Crag) can readily be extirpated by the hand of man, but it is otherwise with the denizens of the sea. It is not easy, indeed, to check the growth of such marine animals, amidst which other types lurk and feed, in turn to become sustenance for the food-fishes, which occasionally, as in the young cod, browse directly on certain zoophytes. Even the gigantic whales engulph many of the little jelly-fishes mingled with other forms on their feeding grounds in the Arctic Ocean.

In this connection the notion which places on the trawl the onus of removing the food of the fishes by rendering the bottom "barren" may briefly be alluded to. Where, when, and how this barrenness has been found is not stated, but its results are said to be as disastrous to the fishes as the destruction of the grass in an enemy's country would be to the flocks and herds. Such writers appear to be unaware of the vast abundance and variety of pelagic beings—from fishes to plants—which own relationship neither to the bottom nor to the locality, but are swept hither and thither to nourish, with cosmopolitan liberality, the fishes of our own and the neighbouring shores. They have overlooked the wealth of life in the sand and mud, which no trawl can seriously disturb. No barren area of this nature in the open sea is known to us, and a consideration of the forces of nature in the sea would show that its barrenness if not wholly hypothetical would be short-lived. A trawl that would simultaneously remove the contents of the water from surface to bottom, sweep the loose forms on the latter as well as those deeply immersed in it—has yet to be invented.

To return from this digression, we observe in other members of the sub-kingdom the same remarkable recuperation and persistence. Though along a limited line of beach the larger anemones, a valued bait for cod, may be removed by the fishermen almost entirely from the tidal rocks, yet the same species flourishes in the neighbouring and deeper waters, and sooner or later the blanks are obliterated. The other large forms in deep water are in no danger, and even crushing under foot on the deck of a ship does not always destroy them, for each piece becomes an independent animal. It is stated that in one of the West Indian Islands a proprietor, wishing to extirpate a colony of gorgeous anemones[1] which attracted

[1] By the kindness of Mr J. E. Duerden, Curator of the Museum of the Institute of Jamaica, I am able to identify this form with *Condylactis passiflora*, Duch. and Mich., a gorgeously coloured anemone. Its column, incapable of retraction, is bright scarlet or orange, either diffuse or in granulations, while distally it is brownish red. The long adhesive and somewhat thick tentacles, which wave gracefully, are also incapable of complete retraction, and are dark brown with minute white granulations, while the tips are bright purple or greenish-yellow, or in partial collapse rich iridescent green.

visitors and so affected the amenity of his grounds, had the holes in the rocks from which they protruded bored by an auger, yet by-and-by recuperation followed, and they expanded their beautiful discs as before.

The anemones which dwell in sand, such as *Peachia* and *Edwardsia*, though occasionally extending in the Channel Islands to the tidal regions, are as a rule beyond the reach of man. Yet the cod pursues the former, it may be after dislodgment by storms, while the stomachs of dabs and flounders are often distended by the latter. Even were it granted that every one of these anemones were swept off an area by storms or by dredges (for trawls do not appear to affect them) nature is not unprepared, for their minute pelagic young are carried everywhere by currents, and even occasionally borne as it were on a graceful carriage by the minute jelly-fishes, to the discs of which they adhere by their tentacles. They then settle on a suitable site and repeople an old area or extend the distribution of the species on a new. Such forms have thus a double protection—firstly by a life in the sand, and, secondly, by a free-swimming larval condition.

The wonderful extension, slow but sure, of coral reefs in tropical regions is another instance of the vast resources of nature—in the development of which man is powerless either to check or modify. The multitudes of free-swimming larvæ of such stocks—far above the needs of coral extension—must form, moreover, an important element in the food-supplies of the ocean in the neighbourhood, whilst some holothurians (sea-cucumbers) and fishes (*Scarus*) feed directly on the coral itself.

In our own waters the dead men's fingers (*Alcyonium*) and sea-pens (*Pennatula* and *Funiculina*) maintain their ground against hook and trawl just as the red coral of commerce does in the Mediterranean and elsewhere.

The capture of marine products for food or for commerce has in some cases been carried on for centuries before science stepped in to ascertain their nature, map out their life-histories, and indicate the true course for legislative interference. The fishery for the red or the precious coral of commerce in this respect agrees with that for the food-fishes. Both had been carried on for centuries before they attracted the earnest attention of the scientific, and both are examples of the long-continued prevalence of error, and, in the case of fishes, even culpable lack of knowledge about a food-supply so important. Indeed, the application of science to the problem of the food-fishes is of much more recent date than that to the coral of commerce, just as if personal adornment, and not practical utility, were of primary importance in the world!

Contrasted with some oceanic forms the red coral is not only represented by limited numbers, comparatively slow growth and moderate powers of increase, but the area of distribution is circumscribed. Moreover, fixed to rocks, stones, shells and dead pieces of coral at the bottom of the sea, it could neither escape the engines of capture, nor, by a pelagic habit, aid in the spread of its larvæ over a wide area. Yet, though hundreds of boats' crews (*Corallini*) annually sweep the ground with heavy bars of wood weighted with stones and fringed with hempen tangles, this valuable treasure of the sea has by no means been extirpated on the very sites where the ancestors of the modern fishermen followed the same pursuit. The pelagic larvæ—perhaps only escaping as the parent-stock is drawn to the surface—plant the species on the same or new areas and defeat man's efforts to destroy it.

Cydippe swimming downwards after engulphing a larval crab (*Zoëa*).

The whole sub-kingdom of the Cœlenterates, therefore, is conspicuous in the pelagic wealth of the sea in every clime, and is a vast and never-failing supply of food to many higher forms, while they (*i.e.* the *Cœlenterates*) in turn occasionally levy a tax on animals as high in the scale as fishes.

For hundreds of years fishermen of various kinds have waged a war of extermination against the common cross-fish or starfish, a form which lives on the bottom, but, like many other marine animals, has free-swimming larvæ (bipinnarians and brachiolarians) which occur in countless swarms during the warm months. Notwithstanding the constant slaughter by liners, mussel-, clam- and oyster-fishermen this species does not seem to be less abundant than it was centuries ago, or within the memory of the oldest inhabitant. While man's pursuit is not stimulated by the value of the starfish in the market or its use as food, yet the injuries it inflicts on the liner by removing his bait, or rendering the fishes unsightly when hooked, and its ravages on oyster-, clam- and mussel-beds suffice to render the annual destruction by man noteworthy, without taking into consideration the loss by storms or by other starfishes (*Solaster*). Though limited areas in shallow water or the tidal regions may be more or less freed from their attacks by constant care, yet taken broadly man has little effect either on their general increase or diminution. Their enormous numbers on certain fishing grounds, on which under favourable circumstances they may be seen closely covering hundreds of square yards of the bottom, besides spreading into deeper waters where they are less visible, is sufficient proof that to-day they are no less numerous than formerly. The old plan of tearing them across the body before returning them to the water only helped to increase their numbers, for each portion of the disc was regenerated and became a complete five-rayed starfish. The expensive measure of collecting them by hand or by other means on oyster- and mussel-beds and placing them on land to dry in the sun is only partially successful, since gulls often carry them back to sea before life is extinct.

The destructive agencies of man have not affected the other members of the starfish-group (Echinoderms) to any appreciable degree. Thus the *Bêche de mer* (Trepang) fishery of Australia and the South Sea Islands seems to show no sign of exhaustion —notwithstanding the eager search for these holothurians as a favourite article of diet for the Chinese. Doubtless the pelagic larvæ, like those of the brittle-stars and sea-urchins, enable them to survive. Indeed it would appear that the echinoderms in which the larvæ are reptant (and not pelagic) are fewer in numbers than those provided with free-swimming young. The group as a whole is important and furnishes food for many invertebrates, fishes and gulls.

Amongst the annelids there are a few forms which have been persistently sought from early times by man for bait, *e.g.* the lobworm (*Arenicola marina*). On limited areas of sand or muddy sand numerous fishermen have almost daily plied their spades, and while, perhaps, the examples may not be so abundant as at first, there is, as a rule, no lack of them on most beaches. The lobworm, indeed, is an instance of a marine form—placed within easy reach of the instruments of capture—which has resisted the attacks of man, probably because a sufficient stock of ripe examples and the very young are covered at all times by the tide. The Palolo of Samoa and Tonga, so much esteemed as food by the natives, and captured in its season with the utmost keenness, is still as abundant as in former generations. This annelid lives amongst the coral reefs of these islands, becomes pelagic for the sake of discharging its eggs into the surrounding water, and is then captured in great quantities for food. Man has persistently taken as many of each species as he could, yet both maintain their abundance.

The enormous numbers of pelagic larval annelids, in successive swarms, which throng the sea throughout the greater part of the year, together with such adults as *Tomopteris* and the *Chætognath, Sagitta,* show in this group alone a great perennial source of nourishment. The demersal adults are everywhere a favourite food of fishes, which find them where man cannot.

In no class is the boundless wealth of the sea in all latitudes more conspicuous than in the crustaceans or crabs. It is necessary, however, to consider them in two groups, viz., the smaller and for the most part pelagic forms, which are not sought by man for food, and the larger, amongst which are the edible kinds. Every ocean and bay, from the Arctic to the Antarctic region, teems with minute crustaceans, chiefly copepods, which at various stages of growth form the food of the younger fishes. It must not be supposed that their small size is a barrier to their being preyed on by the most gigantic inhabitants of the sea, since they form a large proportion of the food of the right whale. In their larval stages (*Nauplii*) they are fitted for the needs of the youngest fishes, such as the cod and the turbot, just after the absorption of the yolk, and they are important links in the cycle of marine life culminating in the fishes. Their ranks are largely augmented by the larval stages of the sea-acorns (*Balani*), which swarm in the inshore waters during the spring months, and afterwards settle in dense multitudes to form calcareous crusts on rocks, boats, stakes and stones in the sea. The copepods occur during every month of the year, and have been found by Dr George Murray and ourselves to feed on diatoms, so that the links from plant to fishes receive another illustration. These and the crowds of the sessile-eyed crustaceans (such as *Parathemisto* in mid-winter) provide an inexhaustible supply of food over which human agency has no influence. Man in a minor way and with great difficulty may protect his boats and ships, the piles of harbours and other places of easy access, from the ravages of the "gribble" (a minute crustacean called *Limnoria*), but he is helpless in the case of floating or submerged timber. Nature almost invariably carries out her own laws, and, as a rule, in the sea these are beyond man's influence.

Another group of crustaceans, viz., the Schizopods, are of importance in most seas, even to a great depth, from their vast abundance at certain seasons, and from their forming a favourite food of valuable fishes such as the herring. For instance in autumn the water near Crail, at the mouth of the Forth, is occasionally crowded with *Thysanoëssa* either alone or accompanied by *Boreophausia* and *Nyctiphanes*—all shrimp-like species of some size, and they may be stranded on the beach in such numbers that it appears to be strewn with long stripes of chaff. Only in the Channel Islands, as at Jersey, are the members of this group (*Mysidæ*) captured by fine nets in masses by men in boats, to be utilized as ground-bait in rod-fishing for mullets[1].

Besides these swimming (pelagic) forms the higher crustaceans inhabiting the bottom, such as the shrimps, prawns, edible and shore-crabs, the hermit-crabs and lobsters, continually send up a series of larvæ to join the free-swimming multitudes, and at a later and larger stage they again pass to the bottom, thus giving the fishes a double opportunity, the smaller seizing them on their upward journey and in their pelagic stage, the older and larger as they descend[2]. The circulation of such larval and post-larval crustaceans in the ocean is thus an important factor in the food of other marine animals.

With regard to the edible crustaceans it cannot be said that man has yet made a notable reduction on the shrimps and prawns so largely captured on many of our shores for food. Where they formerly occurred in great numbers they still prevail; where they are fewer, and where no diminution has been caused by man, they remain as before, without any apparent increase.

It is otherwise with the large slow-growing lobsters and edible crabs, the numbers of the former, especially in our own country and on the shores of Canada, having shown signs of diminution during recent years, from the exertions made to capture them for food. The lobster in its adult condition is chiefly an inshore form, and thus is easily reached by the instruments for capture; while its slowly developing eggs, attached to the abdomen of the female nearly a year, encounter many risks, irrespective of the capture of the adult. It is a species, in short, which readily yields to adverse forces, though at the same time there is no sign of extinction. Its chief safeguard is the pelagic stage of its larva, but as the adults are for the most part inshore forms, and the free-swimming stage

[1] J. Hornell, *Jour. Marine Zool. and Micros.*, II. 1897, p. 90.

[2] Vide *Trawling Report*, 1884, p. 370.

brief, its circumstances are different from those, for instance, of a fish with pelagic eggs.

In this group, therefore, we see that, while the more minute and lower forms are beyond the interference of man, certain of the larger and higher species can be notably diminished. So far as can be ascertained, however, the species found on the bottom of the open sea are as abundant as in former ages, and constitute an important element in the nourishment of the ground-feeding fishes, such as the cod.

The wealth of oceanic life receives a considerable increment from the Polyzoa, which are either tufted plant-like forms or calcareous crusts on stones and sea-weeds. The group, however, is not a conspicuous one in the pelagic fauna, though the larvæ certainly occur in abundance, and for many months of the year. The sea-mats (*Flustræ*) are especially plentiful on some fishing-grounds, and, besides the larvæ which increase the pelagic life, the arborescent tufts give shelter to other forms on which fishes feed.

In the offshore as well as the inshore areas, in shallows and in very great depths, at the surface, in mid-water and on the bottom, the group of the shell-fishes is everywhere disseminated. The vast resources of the sea in this respect cannot be overestimated. For the present purpose the class may be divided into two groups, viz., (1) the pelagic, and (2) those frequenting the bottom.

The pteropods, heteropods, and the pelagic stages of demersal forms—both univalve and bivalve—are universally distributed from the Arctic to the Antarctic seas. Thus *Clione* and *Limacina* form conspicuous elements of the pelagic food of the right whale in the Arctic seas, while the tropical and sub-tropical pteropods are even more numerous. That these and their allies are a favourite food of fishes is well known, though it is less generally understood that even ducks feed on them in the surface-water. Mixed with the purely pelagic mollusks are immense numbers of the larval stages (veligers) of bivalves and univalves, besides sea-slugs and their allies, and they are found during every month of the year, most numerously in spring and summer. The neighbourhood of a mussel-bed, for instance, has a marked influence on the abundance of the pelagic larval mussels, just as the vicinity of a bed of oysters or of clams produces similar effects. In fine weather many of these molluscan larvæ, rising from the bottom, sport at the surface, while by-and-by as they get older they leave the upper regions of the water to descend either to bore in the sand or other medium or take up their habitat on the bottom. The fishes thus, as in the case of the crustaceans, have a double opportunity—first on their rising and again on their descent. The pelagic mollusks, from their enormous numbers and wide distribution, would alone support a great oceanic fauna of predatory animals, and, as they live on microscopic plants and similar minute food, they likewise illustrate the close connection between the two kingdoms. Moreover as food for young fishes they not only give abundance, but afford the necessary variety in dietary.

For ages man has gathered the sedentary and creeping shell-fishes, such as mussels, cockles, and whelks, for food and bait, often without the slightest restriction, as in the case of the whelk and limpet, yet extinction has not ensued, not even in the much abused mussel, which has suffered on the one hand from reckless fishing and on the other from the very varied suppositions of fishermen, mussel-merchants, and politicians. Some years ago an agitation was raised about mussels. Pamphlets and the newspapers of the day kept attention directed to the urgent need for mussel-reform in view of the decrease of the supply. A Committee under Lord Tweedmouth (then Mr Marjoribanks) was appointed by Parliament to investigate the subject. The committee made important recommendations, giving the Fishery Board powers to regulate the various mussel-beds and prevent waste. Very little change, however, has ensued in regard to the distribution or increase of the mussel, and it may be supposed that the agitation in some cases was not disinterested. At any rate, the supply of mussels at this moment is sufficient. The most elementary administration and the bed-system enable this species to maintain its ground. That a populous centre should send companies of cockle-gatherers almost daily to a sandy flat and this for hundreds of years—without exhausting the supply—is evidence that the ways of nature are full of purpose though they may not always be understood.

Man's influence over the preceding mollusks is limited to the inshore, and especially the tidal mussel, cockle, clam, and oyster-beds, to the periwinkle, limpet, and ear-shell between tide-marks, and there it ceases. Even were he to destroy every mussel, cockle, clam, and oyster-bed within his reach, the gaps would be filled up by forms (including those mentioned) beyond his power, and the wealth of molluscan life maintained to succeeding generations. The operations of nature elsewhere are on too vast a scale for his interference. By a single storm she teaches him the inefficiency of dredge, trawl and net, and strews the beach with myriads of shell-fishes of every size, few or none of which are ever disturbed by his operations, and which prove a valuable harvest of bait to the fishermen, and perhaps bring many a predatory food-fish inshore in search of the easily captured spoil. More powerful than the "gribble," the boring shell-fishes (*Xylophaga* and *Teredo*) tunnel their way in submerged timber of all kinds and rapidly disintegrate the masses of wood borne to the sea by great rivers. Man can neither arrest their ravages in unprotected wood throughout the ocean, nor utilize their labours as he desires.

From early times, again, cuttlefishes have been used as food by man, and still more extensively by the larger fishes, such as the bonito, as well as by the whales. No delicacy was valued more highly amongst the ancients on the eastern shores of the Mediterranean than these mollusks. In modern times they have been eagerly sought for bait and so highly prized that as many as possible are captured for this purpose, both on our own and other shores. Some whales, such as the Sperm-whale (Fig. p. 20), live to a large extent on them, so that when this species was more common than it is now the annual consumption must have been great[1]. The majority of the cuttlefishes are, however, pelagic in their adult state, and the young of

[1] It has happened that a sum of £5 or even £10 has been given for a single box of cuttlefishes as bait. The cost of a year's consumption in the case of this whale must be enormous.

almost all are so, and thus they have the protection of a vast area, besides the provision of the inky secretion.

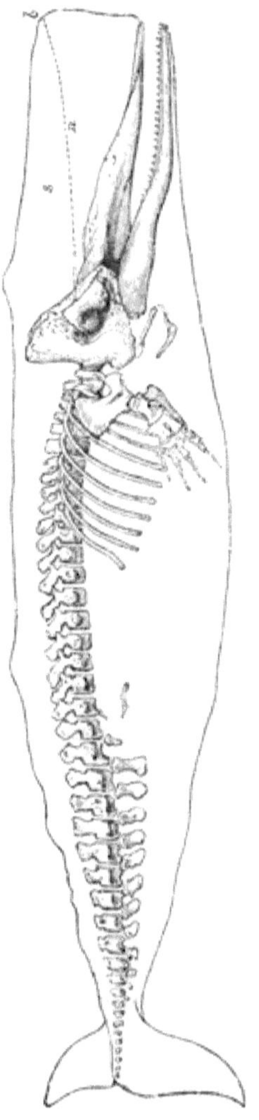

Outline and skeleton of Sperm-whale, which lives largely on cuttlefishes; *s*, region of spermaceti; *n*, nasal passage; *b*, nostril or blow-hole.

It would appear, therefore, that for ages the inhabitants of the sea have largely used the cuttlefishes as food, and to some extent man, both as food and bait, yet these active and generally pelagic forms have not been extirpated. The common species seem to be as abundant as formerly, though their appearance is at all times uncertain. They occur, like so many oceanic forms, suddenly in great numbers on certain grounds, destroying the fishes on the hooks and boldly rising to the surface after their prey as it is drawn up by the fishermen. They also occasionally seriously interfere with the success of the herring fishing (as in 1897) by appearing in great numbers —to the injury of both fishes and nets. In the deeper waters of the Pacific as well as in the other great seas they are abundant, notwithstanding the constant warfare of whales and fishes. This is probably due, as already indicated, as much to the vastness of the area of their distribution, the depths which they frequent, and the provision of the inky cloud, as to the protection of the eggs in tough capsules, by a shell or by enveloping mucus, and the pelagic young. A few, again, have floating eggs, and thus the means for their increase are both varied and extensive as well as beyond the influence of man.

The group of the Urochordates offers few points for remark —except that there is no sign of exhaustion in the common forms within tide-marks and beyond it. The enormous numbers of the pelagic appendicularians and their gelatinous houses is one of the features of the ocean, which in certain places is discoloured by them. Moreover, they feed on microscopic plants, while the smaller fishes and other forms prey on them. The cycle between oceanic plants and fishes has therefore more than one illustration. In this and allied groups the larval form is often more conspicuous than the adult, thus the larva of *Phoronis* abounds in St Andrews Bay from July to September, yet the adult has never been found within it.

We now come to the consideration of by far the most important group in relation to man, viz., the food-fishes of the sea. These comprise the ordinary round fishes, such as the cod and the herring, the various kinds of flat fishes, the skate (belonging to the cartilaginous fishes), and a few others. For ages these have been eagerly followed by generation after generation of men belonging to every nation bordering on the sea. Yet after all these thousands of years can it be said that there are evident signs of the extinction of any modern marine fish?

It has been shown how easy it is to affect the numbers of the larger land-animals, the oceanic mammals and fresh-water fishes, by extensive and long-continued attacks. The problem of the marine food-fishes, however, is less easily solved. The majority of these produce a vast number of minute pelagic eggs, and thus even before the larvæ are born the species is disseminated throughout a great area, it may be, so far as fishing is concerned, of untouched ocean[1]. There is no definite limit, for instance, on either side of the Atlantic beyond which we can say this or that fish does not go. On the contrary, a vast reserve of water more or less unfished by man is always present in the larger seas, and as these are but parts of one great whole which covers three-fourths of the earth's surface, the extermination of any form by man is rendered difficult,—nearly as difficult as if nature had provided insects with aërial eggs, and man had endeavoured to annihilate them.

At first sight it seems almost incredible that such species as the cod, haddock, whiting, plaice, and sole could withstand the vast annual drain caused by the operations of the fishermen of various nations. Yet at this moment all these species in the open seas present as wide a distribution, and, in some, as little diminution in numbers as if the constant persecution of man had not been. Nor must we confine our attention to the ravages of man alone. Whales, seals, sea-birds[2], sharks, dog-fishes, skate, their own and other species of bony fishes, and in their young stages many invertebrates, continually prey on them and have preyed on them for ages. The yearly consumption by the foregoing diverse forms represents an enormous sacrifice of the food-fishes. These and other natural checks are beyond the influence of man, and probably at no period were less powerful than they are now. Man's interference is

[1] Sir J. Murray thinks cooling of the surface destroyed forms having pelagic larvæ, but his grounds are open to doubt.

[2] *E.g.* penguins, gannets, cormorants, guillemots, puffins, gulls, terns, &c.

chiefly confined to a belt within a reasonable distance of land, and to which a constant immigration—active or passive—takes place from more distant waters. It is true that the larger examples of the common species of food-fishes become fewer by persistent fishing, but it cannot be said that, in the case of either round or flat fishes in the majority of the areas, signs of extinction are apparent. The resources of nature, as in most of the invertebrates, would appear to be sufficient for recuperation. The round fishes are a wandering race, moving hither and thither in search of food or for other reasons, and thus their abundance or scarcity, as tested by hook, net, and trawl, is subject to remarkable variations of greater or less duration. The flat fishes, again, by their submersion in the sand, by their activity and by their nomad life in the earlier stages have considerable aids in their preservation. It is probable, however, that it is mainly to the vast number of eggs produced by each species, to their transparency and pelagic nature, and to the existence of boundless reserve-margins of the sea that the food-fishes have been enabled to resist extinction by natural agencies and by the various methods of fishing suggested by the ingenuity of man[1]. Yet the hordes of dog-fishes and the abundance of skate are kept up by forms which have a very few large eggs of slow development, protected either in the body of the parent or in a tough horny capsule. The young fish enters life, therefore, at an advanced stage, and escapes the dangers of minute size and great delicacy.

On the other hand, the vast numbers of the herrings spring from eggs fixed in masses to various structures on the bottom. That the young stages of the cod should seek the tidal margins of the rocks, whereas the young stages of the haddock should keep to deep water, and that the habit in each case should conduce to the prevalence of the species, is another instance of the marvellous prescience of nature.

Even if, in the waters within a reasonable distance of land, fishing were carried to such a degree that it would be no longer profitable to pursue it, it is possible that the adjoining

[1] Enforced close-times, by storms, and in certain parts by ice have also to be considered.

areas and the wonderful powers of increase of the few fishes remaining would by-and-by people the waters as before, because everything in the sea around, including the plenitude of food, so nicely fitted for every stage of growth, would conduce to this end. To those who are annually familiar with square miles of sea (unknown as a spawning-ground) carpeted with myriads of tiny young herrings like fragments of thread, the not uncommon cry of "ruin of the fisheries" seems to need qualification. When, further, in areas supposed to be exhausted,

Sand-eel, a favourite food of fishes both in its adult and young states.

many adult food-fishes are found, whilst the water teems with myriads of pelagic sand-eels[1], flat fishes and other forms, the same necessity for caution holds.

In this group, therefore, as in the majority of the invertebrates, it is apparently beyond man's control either to reduce to vanishing point or greatly to increase the yield of the open sea. The larger forms of such species as the halibut, for instance, may be thinned by constant attacks, but the race continues as before with a resilience and pertinacity none the less sure that they are often doubted and may even be denied. It is a satisfactory proof of the powers of recuperation inherent in the ocean that for ages the British seas, for example, have withstood the almost daily tax of fishermen from both sides of

[1] The richest food of almost all fishes.

the Channel, and the incursions of the eastern boats on the west, in addition to the local population. The wonder, perhaps, is lessened when we consider that for five or six hundred years at least the limited area of the estuary of the Thames has been persistently fished for shrimps by man, and that his nets have simultaneously killed the young soles, plaice and dabs throughout this long period without, up to this date, affecting in any marked manner the prevalence of either crustaceans or fishes. The independence of nature in the sea and man's helplessness are further shown in connection with the swarms of dog-fishes that occasionally occur in the north-west and in the south, and which ruin his captured fishes and nets.

Step by step we have thus briefly passed under review the whole series—from minute floating plants and invertebrates to fishes, and find that they constitute a complex cycle, linked together in intricate bonds, indissoluble by any agency man can devise. Even in the most insignificant group, the operations of nature are on a scale which forbids the possibility of human intervention. Only in certain species exposed by the tide or which frequent shallow water and are easily operated on by man's agency, or are under special conditions in deep water, has the effect been evident. All the rest on a free sea-board appear to resist attempts to reduce to vanishing point or to increase, some becoming for the moment rare or altogether absent, and, just as their doom has been pronounced, reappearing in countless multitudes on the same sites.

The survey of the sea and its inhabitants, therefore, in the main affords no grounds for pessimistic views, but, on the contrary, conduces to reliance on the resources of nature (by which we mean Divine Providence) in this vast area. Generations of men may weave their theories and propound their generalizations, but century after century the oceanic plants and animals maintain their abundance and demonstrate their independence of all artificial regulations. On our shores it may be wise or it may be politic to make such regulations, yet, speaking generally, they seem to have but a feeble effect on the great laws of nature, and especially on the wonderful cycle of life culminating in the food-fishes. The story of these restrictive measures hereafter to be detailed, indeed, shows that they must now be retained on another basis than that afforded by science. We may calculate, as the Duke of Argyll's Commission did, the probable duration of our coal-supply, but the perennial abundance of our marine food-fishes depends on so many factors, each of which has stood every test so well, that, after due precautions and careful experiments, we may, without distrust, look forward to the future.

End-view of a vessel with otter-trawls on board. Aberdeen, August, 1898.

To face p. 26]

CHAPTER II.

Remarks on the Scientific Report to the Royal Commission on Trawling (1884):—

(*a*) Effect of Trawling on the Invertebrate Fauna of the Sea-bottom (forming Fish-food) and Collateral Relations with Pelagic Life.

(*b*) Effects of the Hooks of the Liners on the same Ground.

(*c*) Effects of the Trawl on the Eggs of Fishes, on certain Ground-Fishes, and Very Young Fishes on the Bottom.

The Recommendations of the Royal Commission on Trawling (1883–85), and the method adopted by the Fishery Board for Scotland in carrying them out.

Closure of Areas.

Changes in the Trawling-vessels and their Apparatus.

Changes in the Line-boats and their Apparatus.

The present state of the Line- and the Trawl-Fisheries in relation to the Fishing grounds and the Fishes.

Remarks on the Author's Trawling Report of 1884.

I. *General Remarks.*

Fully fifteen years having elapsed since the Report on Trawling on the eastern shores was presented to the Trawling Commission (composed of the late Earl of Dalhousie, chairman; Right Hon. Edward Marjoribanks, M.P., now Lord Tweedmouth; Prof. Huxley[1]; Mr W. S. Caine, M.P.; and Mr, now Sir T. F. Brady), it appears to be desirable to review the statements contained therein in the light of the information which the impetus given by the Commission has produced. Moreover,

[1] It has been alleged that Prof. Huxley's health prevented him from sharing in the responsibility for the conclusions and recommendations of the Report of the Commission. Cunningham, *Marketable Fishes*, p. 15. He at least read the scientific Report and expressed his approval of it.

this examination of results is all the more necessary, since in 1893 another important body—viz., the Select Committee of the House of Commons on Fisheries, presided over by Mr Marjoribanks, M.P.—issued a new Blue-book containing the finding of the Committee and a mass of evidence.

In criticising this Report on Trawling it is necessary to bear in mind that certain definite instructions were given by the Commission in regard to the hauls of the trawl. These fall under Section 6, and are as follow:—'The results of each haul of the trawl, so far as regards food-fishes, should be carefully registered, in order that positive data may be obtained:

'(*a*) As to the proportional quantity of immature fishes taken at various seasons.

'(*b*) As to the destruction of the spawn of food-fishes.

'(*c*) As to the proportion of live and dead fishes.'

It is important to remember, also, that the choice of ground lay with the trawler in almost every case, and that the most productive ground, so far as could be ascertained, would in all probability be selected.

In the Report of 1884 the fishes were grouped into 'saleable,' 'unsaleable,' and 'young,' the latter term being synonymous with that now in general use, viz. 'immature'—a term, indeed, which was introduced prominently in this Report with precisely the modern meaning. These three heads are well understood, and need cause no ambiguity, since even the fishing community are quite able to understand them—a size-limit, of course, in every case having been considered. To the Royal Commissioners the fact that a young or immature dab was under 7 inches was not of great utility, but the number of such young forms was of the utmost importance in view of the statements then prevalent. Due care was taken to see personally that every example was authenticated, and if any weight is to be attached to the statement that the 'great defect of the Report[1] is that no information whatever is given as to the limit of size

[1] Prof. Ray Lankester, *Sea Fisheries*, Chicago Exhibition, 1893, p. 64. The communication, which appears under Prof. Ray Lankester's name in this publication, is inserted in the *Marketable Fishes* by Mr Cunningham as his own.

dividing the saleable fish from the immature,' there will be little difficulty in remedying it. Besides, it was not the scientific observer who regulated the sizes of the saleable fishes, but fishermen engaged in an industrial pursuit, and who had to bear in mind the demands of the public. Moreover, a fish of a size that was saleable at St Andrews might not be so at Aberdeen, and *vice versa*, though, as a rule, the variation under this head was not great. According to the state of the market, again, fishes—*e.g.*, gurnards—that were saleable at one season were unsaleable at another. As pointed out in the Report, 'It is remarkable that so good a fish should be liable to variation in this respect, and that it should not always be taken to market, even during the height of the herring-season.' Frog-fishes even occasionally found a ready sale in the great central towns of England after the head, skin, and fins were removed; and in the Outer Hebrides dog-fishes formed, and still form, an important item in the crofters' diet-roll, the piles of skins in front of their dwellings being characteristic.

The supposition that because a standard of size is not rigidly in evidence on every occasion, the conclusions will be more or less fallacious, cannot be maintained when dealing with the question of the food-supply. The public do not care whether a fish is mature or immature in the scientific sense of the words, but they are greatly concerned as to whether a fish is saleable or unsaleable. Therefore, after the lapse of a considerable number of years, the author sees no reason to alter the views held during the experiments for the Royal Commission on Trawling, and which have been adopted in dealing with the statistics of the Scotch Fishery Board in the present work. It is true the distinction between the two groups rests on a size-limit, and is by no means a haphazard convenience[1]; but it is unnecessary to dilate on the special features as to maturity or immaturity in an inquiry like the present—however important these may be in other respects. What has to be done is to discriminate between those which are marketable and those which are not. Besides, some writers seem to forget what was clearly stated more than once in the Trawling Report, viz. that the commercial

[1] *Rept. S. F. B.*, 1894, p. 166, and present work, pp. 30, 31.

ships avoided localities where small fishes abounded, whereas the 'Garland' and similar ships did not. They would not dip a second net where these formed the bulk of the haul.

The trawling observations had to do with things as they were and are in the race for marketable fishes, and the exact numbers in each case are reliable.

The statements in the official Report of the Scotch Fishery Board[1] therefore, in regard to the alleged capture of immature fishes by the trawl, may with perfect justice be affirmed of other methods of fishing. For instance, how many mature plaice are caught by the liners in St Andrews Bay? As a rule not one. All are immature, yet the majority are saleable—even to the high sum of 26*s.* or more per half box of 6 stones. The experiments carried on with hook and line on board the 'Garland'[2] can scarcely be taken as average examples. Experience in line-fishing, as it is, proves that young round fishes of all kinds will be captured if the lines are shot amongst them, and perhaps of a smaller size than those caught by the commercial trawl. How often has it happened that the less enterprising liners fishing inshore have filled their boxes only with young haddocks of 5 to 7 inches proceeding inwards from deep water, while their more adventurous neighbours had quite as many boxes of large haddocks from the offshore. It has happened indeed that this capture of the small haddocks day after day has roused the ire of the offshore fishermen, who made a law for themselves, boarded the offenders' boats and flung their small haddocks into the sea, just as in more recent times they did with their neighbours' fishes when they defied the trades-union arrangement, and purchased mussels at a higher rate than was considered reasonable, so as to continue their calling. Very small fishes (*e.g.* of 4 inches) are captured everywhere, that is on the grounds where they occur, by the ordinary methods of fishing.

To take the fishes in the order in which they are mentioned in the Trawling Report, the following sizes formed the lower limit of the saleable fishes:—Skate (including grey, thornback, starry, sandy, &c.), 10–12 in. across the pectorals; herring, 7–8

[1] *8th Ann. Report*, Part III., p. 185.
[2] *Ibid.* p. 189.

in., but those obtained were all much larger; codling (young cod), 8–10 in., but no example so small occurred in the series; haddock, 8–9 in., when so small their price is insignificant—about 1*s.* per box; whiting, 8–9 in.; poor-cod, 7 in.; bib, 6–7 in.; coal-fish, 1 ft.; hake, 1 ft., though seldom seen below 15 in.; ling, 15–20 in.; halibut, 13 in.; sail-fluke, 8 in.; craig-fluke (witch), 7 in.; long-rough dab, 7 in.; turbot, 6–7 in.; brill, 7–8 in.; plaice, 7 in.; dab, 7 in.; lemon-dab, 7 in.; sole, 7 in.; flounder, 7 in., rarely sold; grey gurnard, 9 in.; bream, 9–10 in.; cat-fish or wolf-fish, 15 in., though all those obtained were large. By the term 'saleable,' of course, saleable in the food-market is meant, since much smaller examples of every species might be utilised for manure, either as landed or after preparation in a factory.

In regard to the unsaleable round fishes, the remarks of the Commissioners of 1866 were:—'It has never been alleged that ling, cod, and conger, in which the line fishermen are so largely interested, or mackerel, pilchards, or herrings, upon which the seine- and drift-fishermen depend, are caught by the trawl in an immature and uneatable condition.' 'Whiting and haddocks of small size, thought marketable, are taken by the trawl; but fish of similar dimensions are also captured by the liners, against whom, indeed, the charge of taking immature cod has especially been brought.'

In the Report of 1884 it was stated that 'a considerable number of young cod were present in most of the good hauls, but all were saleable fishes. Quite as many immature cod (codling) were caught by the liners in the same waters; and off the Bell Rock, perhaps, the proportion is even greater.' The same state of matters exists at this moment. On the other hand, the number of very small haddocks caught by the liners, *e.g.*, in 1893, off the east coast of Scotland, far exceeded anything of the kind captured by trawlers. The one mode of fishing was as destructive to these immature forms as the other. The small fishes swarmed on the ground, and were caught in every haul of the liners just as they were swept into the trawl, but many of the smaller forms escaped from the latter through the meshes while they were held fast by the hooks, and so injured that, although they had been returned to the water, it is doubtful if they would have survived.

The remarks made then (1884) on the capture of very young cod and very young haddocks, therefore, remain suitable for to-day, and at this moment (Oct. 6, 1898) numerous boxes of young cod about 10 inches in length come from the lines on the hard ground near Crail; and the same may be said of those on whiting, ling, hake, gurnards, coal-fishes, pollack, bib, and poor-cod.

In the Trawling Report it was stated that large cod and other adult fishes were now seldom caught within the limits of the Bay of St Andrews, and this was in accordance with the evidence then obtainable. The use of anemones as bait, together with the closure of the bay, shows that as many as sixty or eighty good cod are occasionally caught by a single boat, the lines being buoyed and left in the water all night. Some fine congers, which do not appear in the Fishery returns, are also occasionally obtained off the east rocks. Moreover, excellent haddocks are procured in the same area early in the year, and for two years (1893 and 1894) small haddocks abounded. Large green cod also occasionally leap out of the water in pursuit of their prey, and are captured on the beach, while a few pollack are got in the salmon-stake nets or on hooks. It would thus appear that further experience leads to a modification of the statement in the Trawling Report. The increase in numbers captured has been due less to the closure and absence of molestation than to the fixed and extensive lines and special bait.

The closure of the inshore waters—*e.g.*, St Andrews Bay—must have conduced to the prosperity of the turbot and the brill of that neighbourhood, most of the turbot (ranging from 9–11 inches) which formerly were captured by the trawlers (sailing- and steam-) now being unmolested, and reaching the outer waters when of some size. The salmon stake-nets, however, on sandy beaches still prove destructive to many turbot from 5¼ inches upwards. These small examples of this valuable fish are only used as bait for crab-pots. It is true the trawlers sweep the outer waters into which the young turbot and brill pass, but the area is wider, and the size of those captured considerably larger.

No fish formed the subject of greater solicitude in the Trawling Report than the plaice, both from its wide distribution and its great abundance, as well as from the supposed view that this was a form specially destroyed by the trawl, which had cleared out of St Andrews Bay, for example, all the full-grown adults, and left only the smaller forms. It is apparent, therefore, that during the past twelve years such inshore waters have had sufficient time for recuperation—at least to some extent—if these views can be maintained. The results of the trawling-work of the 'Garland' up to 1892 have already been dealt with in this connection[1], so that other observations, and the statistics of fishes captured by the liners in this area, have only to be considered. Without at present going into detail, it is found that comparatively few full-grown plaice are captured in the enclosed waters of St Andrews Bay. Most of the large specimens that have occurred have been either diseased—*e.g.*, blind or emaciated—or injured. An enormous number of immature or half-grown plaice, however, are reared in the area, and are captured by the liners, chiefly with lob-worm, their lines being buoyed and left in the water for such periods as they please, relays of lines being often used. The success with which the local fishermen ply their trade in early spring amongst the plaice is indicated by the fact that a single haul of the lines of a small fishing-boat in February 1894 produced a sum of £9, and that a larger 'catch' was procured by the same boat within the week. The closure of the inshore waters, therefore, while it places the trawl-fishermen at a disadvantage, benefits the line-fishermen, and does not deprive the public altogether of the supply of flat fishes from the enclosed area. It does not, however, produce many large flat fishes, for as these get older they appear to seek the deeper waters outside the limit, either from a natural habit, or as the result of constant interference by man. This habit, indeed, was noticed in the Report when dealing with the question of instituting the closure within the three-mile limit, thus:—

[1] "A Brief Sketch of the Scottish Fisheries," 1882—1892, p. 6.

'The flat fishes, such as turbot, brill, plaice, soles, dabs, and thornback (skate) would certainly be left in comparative security in certain bays, as at St Andrews, the larger only, perhaps, seeking the grounds in the offing.' These larger flat-fishes, many of which are mature (that is, spawning) are captured outside the three-mile limit in great numbers, and thus the supply of ova and young fishes for the inshore waters is affected, for, as previously pointed out, the latter waters depend to a large extent on the former in this respect. Few or no spawning plaice (none within our experience) are ever captured within the bay, though eggs and young in various stages are abundant. It is stated, however, that adult ripe plaice were formerly procured by hook and line off the rocky shore towards the mouth of the bay between Boarhills and Fife-Ness, on hard ground on which no trawl could work. The adult spawning plaice in greater numbers occur in the offshore waters, and, so far as known, there is no passage of these from the outer to the inner area for the purpose of discharging their eggs—as was formerly believed in regard to many fishes. If it had been for the advantage of the eggs and larval plaice that the adults should only spawn close inshore in the shallow water, there is no reason to doubt that such would have been the arrangement. It is apparent, however, that it is otherwise. Before reaching the shallow water of the bays the scattered ova have advanced towards hatching or have hatched, the majority probably in the latter condition, the open water being perhaps better suited for their safety. The yolk-sac of the larval fish is soon absorbed, the symmetrical post-larval condition is reached, by-and-by transformation occurs, and the little fish takes to the bottom, swarms being found in the muddy rock-pools towards the end of April and beginning of May. The life-history of this species would seem to show that—in dealing artificially with the eggs and larvæ—the most natural method is to place the larval fishes, just before the yolk-sac is absorbed, some distance from shore. They are more or less transparent, and will escape many of the dangers they run in such waters, and, before being carried close inshore, will either be transformed or about to be transformed, and more capable of escaping, by their own exertions, from their enemies. If the larvæ are placed in the sea close to a rocky beach or stretch of tidal sand or gravel, it is possible that many would be stranded by the tide. Therefore, though the observation that the young plaice (with eyes now on the right side) abound in spring in the shallow rock-pools and elsewhere is perfectly correct, it is no argument for placing the larval fishes in their neighbourhood, when in a truly pelagic condition. In the same way the spawning ling are found far from the inshore waters, their minute eggs being hatched in the open ocean, and the young stages passed long before reaching the margin of low water. The ling has not, indeed, been found in inshore waters till it reaches about 3 inches ($3\frac{1}{2}$) in length, and then in very limited numbers. It is more frequently secured when from 6 to 8 inches in length—at extreme low water at the margins of the rocks. As it gets larger it seeks the offshore, and thus, as in the plaice, there is a double migration—the wafting of the eggs, larval and young fishes shorewards, and the return of the adolescent and the larger forms seawards. A similar life-history appears to be present in many of the food-fishes—*e.g.*, the turbot, brill, and halibut, though in the case of the dab, long-rough dab, and some others, there are marked exceptions, as pointed out in the Trawling Report. Thus, 'the large proportion of immature dabs found 15 miles off St Abb's Head is interesting, and shows that such are not confined to shallow bays like that of St Andrews. Moreover, the occurrence of relatively small specimens at this and even greater distances from land would raise a doubt as to whether all such young forms have been reared on a sandy beach inshore[1].' Since the foregoing was written, opportunities, by aid of the 'Garland,' for using the special trawl-like tow-net and the mid-water net near and at the bottom on the grounds 15 to 20 miles south-east of the Island of May, have been afforded, and great numbers of larval, post-larval, and young dabs, long-rough dabs, and other forms have been obtained, thus confirming the opinion formerly expressed. Moreover, the trawling work of the 'Garland' on its various stations from the Moray Firth to the Forth bears out

[1] *Report Roy. Com. on Trawling*, p. 361.

the same conclusion. Again, the deeper water is the home of the post-larval frog-fish, even the pelagic eggs being rather uncommon near shore. The adolescent and adults, on the other hand, are frequent in shallow sandy bays like St Andrews.

It is apparent, from certain remarks in the preceding paragraph, that it is a mistake to say that the trawl alone can capture flat fishes. If the bait be suitable, the lines are tolerably effective in regard to plaice, lemon-dabs, dabs, and flounders. Again, halibut-fishing (by hook) is the most productive method off the coasts of Iceland, Faröe, and elsewhere, and even the turbot and the sole are occasionally caught by the liners.

a. *Effect of Trawling on the Invertebrate Fauna of the Sea-bottom (forming Fish-food), and Collateral Relations with Pelagic Life.*

The value of the bottom-fauna, in regard to the sustenance of the food-fishes, has been fully recognised by all zoologists. In the Trawling Report for 1884 it was stated[1], 'There cannot be a doubt about the importance of maintaining the invertebrate fauna of these parts (Forth) in a flourishing condition, since upon this many of the food-fishes depend for much of their nourishment; indeed, both adult and young fishes could hardly exist without such, notwithstanding the abundance of herring and other pelagic food at some seasons[2].'

In considering the effect of the trawl on the sea-bottom, it must be borne in mind that while many sponges, zoophytes, star-fishes, crabs, and shell-fishes are, in their adult state, inhabitants of the bottom, their larvæ and young are pelagic, that is, free-swimming, and quite beyond the reach of injury, so that were the majority of their parents killed and the sea-bottom rendered barren[3] (of which, apparently, there is an

[1] "A Brief Sketch of the Scottish Fisheries," 1882—1892, p. 6.

[2] *Op. cit.*, p. 370.

[3] *Vide* an interesting Essay on this subject by Mr Anderson Smith, *Trans. Highland and Agricultural Society*, 1890, p. 45.

absence of proof), swarms of the young would settle on the sites thus bereft of their predecessors. This interchange between the surrounding water and the bottom was pointed out in the Trawling Report, thus:—'The sedentary fauna of any such ground'—referring to fishing-banks—'brought up in the trawl, does not, however, give the whole explanation, for it has to be remembered that almost throughout the entire year a constant succession of eggs or young forms is given off by the inhabitants' (on the bottom), 'while there is another and less minute increment derived from the older forms which are forsaking pelagic life to settle on the banks[1].' It is the vast abundance of the pelagic fauna (which often has no connection with the particular ground examined), for instance, that is so important in regard to the food of the herring.

Again, the forms immersed in the sand are, for the most part, free from serious injury. This was pointed out, speaking of certain annelids, in the Trawling Report, as follows:—'The trawl has little or no effect on such as the foregoing, for their tubes in general are buried in the sand, their lives being passed in boring through it—for food and for shelter. So difficult is it to interfere with annelids sunk in the sand that one may sweep over a sandy surface, in which rare forms lurk, again and again, without discovering their presence. Even if a dredge is dipped, perhaps, only a head with its tentacles will reward the haul. Yet the sand may teem with them from 5 fathoms shorewards, and every spadeful—at extreme low water—may produce several. A single storm would appear to inflict greater destruction on these than it is possible to do by trawling[2].'

A careful summary of the effects of the trawl on the sea-bottom and its fauna was given at the end of the chapter on the 'Fauna of the Trawling-Grounds in relation to Food-Fishes[3],' and, with all the experience since gained, it is doubtful if any change can conscientiously be made. The only fact of moment to be stated is that it is several years

[1] *Report, Royal Commission on Trawling*, 1885, p. 370.
[2] *Ibid.*, 1885, p. 368.
[3] *Ibid.*, p. 370.

since the West Sands at St Andrews have been so extensively covered with marine organisms in débris, as, for instance, in the spring of 1857, when the misty air was alive with hooded crows and gulls that fed on the stranded animals. Smaller numbers have been seen, but not the vast numbers of sea-mice, sea-urchins, star-fishes, shell-fishes, zoophytes, crabs, and other forms.

Besides the evidence (and there is much) of the capture and injury of the invertebrates in the trawl, and the consequent removal of part of the food of the fishes at the bottom, it is necessary to show that by constant trawling the sea-bottom is rendered destitute of nourishment for the food-fishes. Nature has provided in the case of such forms a remarkable power of recuperation, and a vitality that renders complete destruction difficult. Thus, the crushing and division of sponges is not followed by the death of all the fragments, and each of those which survives is capable of flourishing as an independent organism (not to allude to the liberation of ova which may happen to be present). The ordinary zoophytes are, as a rule, little affected by being carried on board ship, for, though in the more delicate, such as *Tubularia*, the soft polypites may fall off, they are reproduced when returned to the water. The very general provision of free-swimming buds which bear the reproductive elements is also a complete safeguard in the majority of this group, as already stated. Whence do the immense multitudes of the zoophyte (*Obelia*), which coats with a feathery forest every rope and buoy of the salmon stake-nets, come? Certainly not from the bottom, but from the myriads of minute pelagic jelly-fishes. Each half of a bisected anemone becomes an independent animal. Again, the sand-dwelling forms, such as *Peachia* and *Edwardsia*, the latter occasionally eaten by flounders and dabs, have pelagic young which are swept in every direction by the currents, or sometimes are carried hither and thither attached to the disks of the swimming jelly-fishes. These forms are, indeed, doubly protected, since they live in the sand, and are seldom or never captured by the trawl, while their young are wholly pelagic. 'The masses of dead men's fingers (*Alcyonium*), though occasionally crushed, would, for the most part, survive after their presence on deck[1].' The sea-pen (*Pennatula*), unless seriously crushed, readily survives—whether captured by trawl or hook, and would have no difficulty in again taking up its position on the bottom. 'More decided injury is inflicted in many cases on the members of the star-fish group, which form no inconsiderable part of the food of the cod, haddock, and flat-fishes. Their brittle nature can ill withstand the rude trials of the trawl, and still less the trampling on the deck of the trawler, when the fishes are packed. All suffer more or less, the majority seriously; the forms most liable being brittle-stars, sea-urchins, and heart-urchins[2].' Such was the view expressed in the *Trawling Report*, and, though less pronounced than some more recent opinions, it appears to state the case not unfavourably for the opponents of the trawl. Thus, the sea-cucumbers, or Holothurians, often brought up from the bottom amidst stones and sea-weeds, are, for the most part, uninjured; and, since they can voluntarily eject their entire alimentary system, and, as the patient and persevering Sir J. G. Dalyell showed, reproduce it without serious inconvenience in about three months, they are not likely to suffer from the hands of the trawler. The cod is a more exacting marauder of the bottom in regard to the smaller Holothurians, since they are chiefly found in its stomach, and not in the trawl. The 'dreg' of the Zetlandic fisherman, as he searches for 'yoags' (horse-mussels) for bait, is also a more effective engine for the capture of the great 'sea-puddings,' as these Holothurians are called. While the brittle-stars are mutilated, they are by no means in all cases killed. The disk produces new arms where they have been broken off, and even an injured disk is repaired. While many of the sand-stars are entangled in the trawl, the majority escape by being imbedded in the sand, as can easily be shown by using a dredge on the same ground over which many a trawl has passed. The sea-urchins suffer considerably by the trawl in certain regions, as, for instance, in the outer parts of the estuary of the Tay, and it is doubtful if many of those

[1] *Trawling Report*, 1884, p. 370.
[2] *Ibid.*, 1884, p. 370.

injured would survive. On the other hand, many fine examples frequent the rocky borders where no trawl can touch them. The delicate heart-urchins are, for the most part, ruined in the trawl, yet the habitat of many is deeply buried in the sand beyond the reach of 'sole'-rope or iron trawl-heads. Then, again, how many liners, how many mussel-, clam-, and oyster-farmers, would gladly subsidise the trawler to remove the swarms of common cross-fishes, whose multitudes form, for acres, a carpet on the bed of the ocean!

Not a few annelids, such as sea-mice, nereids, 'scale-backs,' serpulids, and nemerteans are entangled in the ground-rope and the net of the trawl, or in old shells and other débris from the bottom; but the injury to this group, as formerly mentioned, is not great, for many reach the sea alive amongst the débris, and regenerate lost parts or discharge their ova into the surrounding water, or the larvæ swim from under the scales or from the sacs. It is an interesting fact in this connection that certain annelids at the breeding season undergo a change of form, leave the rocks, cavities under stones, or other places of shelter, and swim freely in the water (that is, become pelagic), discharging their reproductive elements when thus *en voyage*. It is during these pelagic periods that the huge *Alitta virens* (a worm reaching occasionally about 3 feet, and a valuable bait for fishes) is thrown by storms in great numbers on the sands, even before the function for which nature ordained the pelagic period is performed. In the same way, Mr Thomas Scott, and Mr Duthie, recently found the inshore water at Castlebay, Barra, swarming with the sexual forms of a *Nereis*. An examination of the stomachs of food-fishes shows that such a provision as the foregoing is duly taken advantage of by them, and it is well, since the annelids are destined to perish after the escape of the reproductive elements. This subject was specially treated of many years ago[1], and in the recent Reports of the Fishery Board further investigations have been made by Mr Ramsay Smith[2]. Some annelids, again, are purely pelagic,

[1] *Invertebrate Fauna and Fishes of St Andrews*, p. 101, *et seq.*, and also Dr Day in *Literature, Fisheries Exhib.* 1883.

[2] *Vide Annual Report of the Fishery Board for Scotland*, 1895.

like *Alciope* and *Tomopteris*, the latter being one of the most striking features of the tow-nets, both inshore and offshore, and frequently in great abundance. Moreover, like the masses of *Sagitta*, it is eaten by many food-fishes, and is another evidence that the question is viewed only from one side, when all the wealth of the pelagic fauna and flora, especially the multitudes that are independent of any particular inshore area, or of any land at all, is overlooked.

In 1884, after a careful and extensive inquiry, the effect of the trawl on the crab-tribe or crustaceans was given as follows:—'Like the star-fishes, the crustaceans are evidently damaged less by the effects of the trawl than by the feet of the men in gathering the fishes.'...'Many hermit-crabs, sea-acorns (*Balani*), and *Galatheæ* are returned to the sea alive. No injury to soft crabs was observed, and even so slender a form as the northern stone-crab, besides others, have been brought on deck in good condition[1].' The most prominent form referred to was the Norway lobster, which occurs in great numbers off the Forth, and a caution was given that care should be taken to return it to the sea alive, for it is chiefly injured on deck, in selecting the fishes. This is a very important element in the diet of the cod in the neighbourhood, and every measure should be taken for its preservation. The muscular parts of the abdomen form excellent food, but they are seldom brought to market. After hatching, under the abdomen of the female, the young of this lobster is pelagic, and is often found in masses amongst jelly-fishes off the Isle of May, forming a rich nourishment for the smaller fishes. Not a single lobster was observed in the trawl, and only from the inshore grounds, north of Aberdeen, were a few edible crabs obtained. The condition of the lobster- and crab-fisheries near St Andrews, Dunbar and elsewhere has improved within recent years, but how far this is due to the security from molestation which the crab-traps now have, and which has caused an increase in their number, or to other circumstances, is unknown. Taken generally, however, in Scotland, and so far as the statistics go, it is seen that there is a decrease in the quinquennial period 1888–1892 of 107, 840

[1] *Op. cit.*, p. 370.

lobsters, and 2013 crabs,—from that in 1883–1887. The decrease, however, seems to be in no way connected with the trawling industry. An increase both in quantity and value has since occurred. Trawling close inshore with a small-meshed net (naturalists') often produces large numbers of swimming crabs, but such is a rare occurrence on board an ordinary trawler.

While, therefore, in trawling the injury to the crustaceans inhabiting the bottom is considerable, it has to be borne in mind that the pelagic crustacean fauna is one of the most marked features in the ocean, from the north to the south pole, and some of its members, *e.g.*, barnacles attached to floating timber, sessile-eyed crustaceans, schizopods, and thysanopods are large enough to be the food of haddocks and herrings. Moreover, vast swarms of the smaller copepods nourish the younger food-fishes, and other types, again, are eaten by the larger fishes. Near the mouth of the Forth, at certain seasons (viz., in autumn), the water near the 'Hairst' at Crail is almost thickened by *Thysanoessa*. The tow-nets, under these circumstances, soon become filled with their masses. Some, however, may be disposed to treat the pelagic crustacean food, in connection with the nourishment of the fishes, with indifference, deeming such small animals of little moment in contrast with the more substantial denizens of the bottom. Irrespective of the circumstance that where they abound other and larger forms are in their wake, the fact that so gigantic an animal as the right, or whalebone-whale of commerce, lives solely upon such pelagic food in the Arctic seas, is sufficient to show how important an element this floating or swimming fauna is in regard to the welfare of the fisheries. The larger part of the masses in the tow-net, kindly used in 1893 by Dr Allan of Glasgow on board the whaling-ship 'Aurora' of Dundee, consisted of the little crustaceans (copepods) above-mentioned, and this in the actual food-line of the whale as it swam to and fro in the water near an ice-pack. In our own waters, in addition to what is mentioned above, the sea-acorns, which so extensively cover rocks, stones, shells, and other structures within tide-mark, as well in deeper water (not to speak of

Baiting crab-pots on St Andrews "slip," June, 1896.

To face p. 43]

those adherent to the skins of various whales), send off a multitude of free-swimming young which often crowd the inshore water, and extend far beyond before settling down on every available surface. Many other instances of the plenitude of crustacean pelagic life, and its adaptation for the nourishment of fishes, might be given, but sufficient has been cited to show that, besides the crustaceans of the bottom, those frequenting the water itself must be considered.

The ascidians or 'sea-squirts' of the bottom are occasionally brought up in the trawl attached to shells, stones, and sea-weeds; and such forms, along with pieces of 'sea-mat,' are not unfrequent in the stomachs of cod and haddock. They are usually sent overboard in a condition by no means unfavourable for existence, though it has to be stated that numerous pieces of adherent sea-mat are often brought to shore on the ground-rope and meshes of the trawl from certain areas. The young ascidians are free-swimming (tadpole-like), and escape interference till they settle on shells, stones, and sea-weeds. An interesting species in this group (*Oikopleura* or *Appendicularia*) is pelagic throughout life, and often occurs with its gelatinous 'houses' in dense multitudes in our inshore waters. It lives upon microscopic plants and similar structures in the water, while the smaller fishes and other forms prey on it.

The last group of the invertebrates liable to injury by the trawl is that of the shell-fishes and cuttle-fishes (Mollusca). In 1884 the opinion expressed was that, 'amongst the mollusks, the nudibranchs suffer a little on deck, but the cuttle-fishes are more delicate, the majority being almost lifeless on removal from the net. The horse-mussels are occasionally fractured, but the whelks are uninjured.' With the exception of water-logged wood bored by the ship-worm, horse-mussels, and whelks of various kinds, the majority of the shells brought on board were old and empty, either covered with growths of various kinds, such as zoophytes, or harbouring star-fishes and annelids in their interstices. As a rule, the horse-mussels were uninjured, and were consigned in safety to the water with the débris. On some of these such delicate organisms as the spawn of the Norwegian whelk (*Fusus norvegicus*), with the contained embryos in a thin capsule nearly an inch in diameter, were in perfect condition, and are now in the University Museum. These mussels occur in deep water, and often form large masses bound together by the threads of the byssus or 'beard,' and are quite as frequently drawn up by the liners on their hooks on their particular ground. Moreover, some of the liners for years used to bring quantities close inshore and deposit them off the rocks, thinking to create a bed of horse-mussels, but all disappeared. Not a trace remains of the many tons of horse-mussels thus transplanted, except an occasional and solitary small example in a chink of the rocks, perhaps an inshore and last surviving descendant of the transplanted shell-fishes. The living whelks were generally entire, those (*Fusus*) having the rare anemone (*Hormathia* the 'necklet') adhering externally, being in perfect condition, just as the much more delicate dead *Natica*, overgrown with sponge, and tenanted by a hermit crab, often came up uninjured. Masses of the spawn both of the great whelk and *Fusus* are frequently brought up by the trawl, but much of these is uninjured on again reaching the water. Even on great stretches of sand, in which many shell-fishes abound, comparatively few bivalves are interfered with, since they are buried more or less under the surface, and afford little hold to the ground-rope. The molluscan fauna of muddy ground is also inconspicuous in regard to injury, the smaller forms, which are sometimes numerous, escaping entirely. A single severe storm dislodges those on sandy ground more surely and extensively than years of continual trawling. The spawn of the whelks, and that of the nudibranchs and cuttle-fishes, is more likely to suffer, yet only to a limited extent, since all is again consigned to the water; and even a somewhat lengthened exposure on deck is not fatal, if a little moisture be present, since many can be hatched after the arrival of the ship in port. In these remarks trawling over a mussel-bed, a clam-bed, or over an oyster-bed, is not considered, since it is strictly and rightly prohibited. On the whole, then, the shell-fishes do not suffer conspicuous injury by the use of the ordinary trawl, a fact sufficiently plain at St Andrews, where the larger mollusks were eagerly sought for bait by the fishermen using trawls. It was only after a storm that a few were occasionally procured in a trawl, but they never failed to find multitudes strewn on the beach at the West Sands after a severe storm, of which occurrences, indeed, zoologists have frequent personal experiences.

Many cuttle-fishes, again, are captured by the trawl, and, as above mentioned, as a rule, are killed. But these are carefully preserved for sale, and in certain cases form no inconsiderable item in the proceeds of the fishing, for a sum of money, varying from £1 to £5 or even double the amount, is paid for each box of this valued bait. Further, the cuttle-fishes caught by the trawl are chiefly *Loligo* and *Ommastrephes*—squids as they are usually called. Now, these are pelagic or free-swimming cuttle-fishes that do not necessarily fall under the bottom-fauna, though, when captured, they are probably seeking their prey there, or in the stratum just above it. The occurrence of *Eledone* (a ground-form) in the trawl is less common than the foregoing. Moreover, the squids are mollusks particularly obnoxious in the active condition to liners, so much so that advice has been asked as to how to get rid of them[1]. They occur in such numbers now and then that the fishermen despair of their catches, for the cuttle-fishes so disfigure the haddocks and other fishes fixed on the hooks—by devouring the muscles behind the head—as to render them unsaleable. So eager are they, indeed, that they sometimes follow the hooked fishes to the surface, as the lines are hauled, and are captured by a hand-net. Opinions, therefore, might differ as to the disadvantages of thinning this group of mollusks, which, when full-grown, are as much destined for the nourishment of the whales as the food-fishes, though they are also eaten by larger examples of the latter (*e.g.*, the cod), and in their younger stages by many others.

The case of the molluscan fauna of the bottom as food for fishes in relation to the action of the trawl, however, cannot be considered without the pelagic or free-swimming representatives of the same group on the particular ground. Almost all the shell-fishes living on the bottom send off pelagic young,

[1] *Vide Fourth Annual Report of the Fishery Board for Scotland*, p. 204.

which crowd, throughout the greater part of the year, the region over the bottom, and ascend to the surface. Swarms of purely pelagic forms, such as *Spirialis*, which, small though it be, is often eaten by ducks at the surface, and more abundantly by many young fishes near the bottom, are mingled with these and the larval nudibranchs, and occasionally even with *Clione*. The latter does not attain the bulk of the Arctic examples, but is by no means infrequent at certain seasons. Finally, the proximity of a mussel-bed, or the presence of an extensive coating of small mussels on the rocks, fills the water (June and July) with myriads of pelagic mussels, which by-and-by adhere to everything affording a surface, and form a favourite food of young fishes, and even of some of the adults.

Irrespective of its direct relation to fish-food, this unceasing wealth of pelagic life has a close connection with the sustenance of the bottom-fauna. Every invertebrate group, mentioned in the foregoing paragraphs, feeds on the pelagic fauna or its débris; for, even the highest—viz., the cuttle-fishes—frequently devour the pelagic fishes. The whole system forms a wonderful cycle, and is far more important, from a fisheries' point of view, than at first sight appears.

In returning the contents of the trawl to the water, many forms are doubtless devoured by fishes in their descent to the bottom, just as the gulls which follow the herring-boats in the western lochs, or the wake of a trawler on the eastern coasts, swoop on the offal thrown overboard, or on the injured fishes which escape from the net at the surface when hauling. The same remark applies to the invertebrates thrown overboard by the liners.

b. Effects of the Hooks of the Liners on the same Ground.

In considering the effect of the trawl on the animals frequenting the bottom of the so-called fishing banks, it must be remembered that for ages the hooks of the liners have brought from the same banks representatives of every group of invertebrate animals, and that they are not always replaced in the sea alive. To the hooks of the liners, and also to their ready courtesy, most of the museums in this and other countries owe much. Some of the finest sponges in British waters, and the beautiful Venus's flower-baskets and glass-rope sponges abroad have been procured in this way. A constant and large supply of hydroid zoophytes ('sea-trees') is almost daily brought on shore from the deeper water. Some of the largest and finest anemones—studded, it may be, on the mandible of a small rorquhal, on a piece of submerged wood, on flat or other stones, on shells, or even on the thigh-bone of an unfortunate sailor—are similarly procured, along with stony corals and other coral-like structures (Polyzoa). If the trawler is accredited with the destruction of the great sea-pen (*Funiculina*)[1] of the western waters, what is to be said of its ruddy ally (*Pennatula*), the 'pink' of the eastern fishermen? Almost every hook for considerable portions of lines on certain grounds bringing up its example, and this not by any action on the part of this pretty sea-pen, but simply by the force of the tide on the line as it drags the hooks over the surface of the soft ground. Several jars have been filled in a single trip with these from such sources, and yet no scarcity of them exists, nor is it hinted that the cod directly, or the haddock indirectly, will be robbed of its food. In the same way fine masses of 'dead men's fingers' (*Alcyonium*), the slender sea-pen of the Forth (*Virgularia*), a rarer and larger type new to this country, and other forms, are captured by the liners.

Before and since the days of Edward Forbes the liner has been the mainstay for many rare star-fishes, and those who, like the genial naturalist just mentioned, have eagerly waited in the dim morning light—with the ready pail of fresh water, or, still better, with the jar of strong spirit—for the advent of the brittle *Luidia* on board will appreciate the services of the

[1] The late lamented Professor Milnes Marshall thought that the cod had a particular fancy for these great sea-pens, biting the tips as it swam amidst the phosphorescent stems, and leaving many in a truncated condition.

fishermen in this respect. The daily captures of the common cross-fishes by the liners amount to a great annual total, yet such is unavoidable, and, indeed, the species is a pest to both the liner and the shell-fish farmer. As already said, it is the liner who procures many of the smaller holothurians, and who first made us acquainted with the rarer heart-urchins and sea-urchins (*e.g.*, the 'piper').

In the group of the worms, one of the most gigantic Nemerteans (3 feet long and nearly an inch broad) has hitherto only come from the liner, and even the black line-worm is occasionally caught on the hooks. In the débris from deep-sea fishing-boats many of the rarer bristled annelids have been procured, and serpulids, the coral-like masses of *Filigrana*, and many others, come from the same method of fishing. It is an interesting fact, and a criticism on man's influence on the inhabitants of the ocean, that neither liner nor trawler has ever captured the characteristic spoonworm (*Echiurus pallasii*) of St Andrews, and that storms alone toss them in multitudes on the beach. Food-fishes, however, find them out, and feed on them in their haunts.

If we regard crab- and lobster-fishing as a branch of the trade of the liner—and it apparently has little connection in any respect with the trawler—the effect of other instruments than the trawl in reducing the number of these animals is conspicuous. The larger forms by-and-by become rare, and all become fewer. There is no proof that the trawl ever seriously affects either species, but its use, close to rocky borders in bays, may occasionally in former days have interfered with the traps. The hooks of the liner now and then capture both species, and furnish many hermit-crabs in shells, Norway lobsters, some of the rarer long-tailed forms (*e.g.*, *Munida*), and many of the short-tailed crabs (*e.g.*, *Atelecyclus* and *Eurynome*). The hooks, likewise, bring up branches of submerged trees, coated with large sea-acorns, and affording shelter around the feathery tufts of zoophytes to many small sessile-eyed crabs.

Fine masses of sea-mats and bunches of sea-grapes (Ascidians) are very common on board the liners, the hooks penetrating the tough tests of the latter (Ascidians) as the lines are carried by the currents. Only from the liners are the curious and rather rare 'sandy-nipples' (*Pelonaia*) procured—another of the remarkable forms characteristic of St Andrews. Hundreds are sometimes brought on shore, the drifting hooks having penetrated the thinner superior part of the test as the creature was plying its respiratory and nutritive currents on the bottom. None have yet been obtained from the trawl, for the ground is rough. The multitudes of dead shells brought up by the liners have been one of the richest fields for the encrusting forms of the Polyzoa—such as *Lepralia*, the foliaceous—as *Eschara* and *Retepora*, and the more or less ramified *Cellepora* and *Serialaria*.

When we come to the group of the shell-fishes or mollusks, the great variety that have been obtained from the hooks of the liners for hundreds of years is noteworthy. No collection of marine shells, worthy of the name, and made long before trawling was introduced into Scottish waters, exists—but is a standing proof of the fact that the hooks are almost as effective in furnishing rare specimens as the trawl. Some of the smaller forms, indeed, are more surely obtained in this way than by the trawl. Many are found in the débris attached to the masses of horse-mussels so frequently brought up from the bottom, others are fixed to dead valves of the larger shells, or have their siphons, foot or other parts, pierced by the hooks. On the zoophytes brought up by the lines, such naked forms as *Doto*, and, on *Alcyonium*, *Tritonia plebeia* are found in numbers, while the larger *Tritonia Hombergii*, "a sea lemon," is occasionally pierced by a hook as it drags over the bottom. The active cuttle-fishes do not seem to be frequently caught by the hooks, though, when attacking the fishes on the lines, or interfering with bait, an arm or other part is now and then fixed. In contrast with the trawl, however, few cuttle-fishes are captured by the lines.

It is by no means to be supposed from the foregoing remarks that any blame is to be attached to the liner for interfering with the bottom-fauna. The notion would be unwarrantable. In the course of his operations the forms indicated have been unavoidably hooked, and he brings them on shore or throws them overboard without giving much heed to them, unless requested to do so. No marine zoologist of note would consider, on this head, that the liner was ruining the food of fishes, or to any extent affecting the welfare of the fisheries. On the contrary, his ready courtesy and keen observation have been frequently of the greatest service to the naturalist, who has been in the past, and is in the present, greatly indebted to him for many rarities. Yet the fact remains that the liner likewise captures numbers of every group whose wholesale destruction is by some attributed to the trawler. A calm survey of the situation, therefore, does not lend support to the notion that the trawl, as ordinarily employed in sea-fishing, is the only destroyer of the invertebrate animals of the bottom; and, further, experience does not demonstrate that the sea-bottom in any known region has been, by the use of such line or trawl, so seriously impoverished as to be unable to support fish-life.

c. *Effects of the Trawl on the Eggs of Fishes, on certain Ground-Fishes, and very Young Fishes on the Bottom.*

In the Report of 1884 it was stated that 'no feature was more remarkable in the inquiry than the rarity of fish-spawn (eggs) in the trawl—notwithstanding the careful search for such on every occasion.' A few zoophytes, with clusters of adherent eggs—then considered to be those of the herring—were all that could be found. Accordingly, it was reported 'that the trawl is almost innocuous so far as the ova of fishes is concerned;' and doubt was expressed, even if it passed over masses of the eggs of the herring, whether injury would always take place.

It has frequently been stated that the use of the trawl on the spawning-grounds of the herring drives away the herring and injures the spawn and spawning fishes. No details, however, are given as to whether line-trawling or beam-trawling is referred to, nor are the facts dealing with the destruction of the spawn and the lesions of the spawning fishes presented in a tangible form. The herring is a fish sometimes exceedingly tenacious of purpose, and a thousand intervening boats with their nets, as at Peterhead, will not prevent it from breaking through and spawning inshore. Again, it has been known to leave old spawning-grounds, *e.g.*, the 'Old Hake' off the coast of Fife, upon which no trawl ever descended, or indeed could descend with safety. Many reasons have been given for such changes—from the breaking up of the shoal by the boats at sea, to the firing of big guns—but there is a lack of definition. In the Trawling Report of 1884, it was recommended that trawlers (*i.e.*, beam-trawlers) should avoid shoals of herring, and the same suggestion holds to-day.

Since 1884, no opportunity of observing the effect of a trawl on ground covered by the spawn of the herring has occurred. The boats employed in herring-fishing in winter, however, frequently bring to port masses of the spawn of the herring on their decks, and little difficulty is found in hatching these eggs in the Laboratory, even after sixteen hours' exposure on deck. Thus, even if the eggs of the herring were brought on board during the operations of the trawl, there are good grounds for believing that many would survive after being replaced in the water. In St Andrews Bay the local trawlers formerly brought to land pieces of seaweed, zoophytes, and similar structures to which the eggs of Montagu's sucker and other forms adhered. These were all readily hatched in the Laboratory. The same eggs are frequently brought up on the hooks of the liners. Further, about the middle of January 1886, one of these local trawlers brought a huge adherent mass of large eggs amongst mud to the harbour—thinking it was the spawn of the salmon. After lying on deck a considerable time, the attendant removed the mass from the mud and débris, rather rudely tore it in several pieces, and placed them under sea-water—some of the advanced embryos escaping in the process. The majority of these eggs were hatched, and the larval fishes were vigorous, and enabled an account to be given of the development of the wolf-fish ('cat-fish' of the fishermen), to which it was found the eggs pertained. At least once since the same eggs have been brought up in the trawl (from the Forth) about the period of hatching. It was rare to find, even in such sandy bays as St Andrews, that the egg-'purses' of the thornback or other forms were brought in by the trawl, though many containing embryos are stranded on the West Sands after storms in October, and especially in November. Such eggs are at least as frequently procured on board the liners, *e.g.*, those of the grey, starry, and sandy rays. Before the spawning of the sand-eel was fully elucidated[1], it was often a matter of conjecture as to how the ova escaped the small trawls on inshore ground; but now the eggs are known to be adhesive—probably to the sand, and thus avoid interference.

The ordinary trawl, again, seldom or never retains such active ground-fishes as the smaller rocklings, the gobies, the gunnel, the eel, and the sand-eel, while the glutinous hag and the lamprey are almost unknown in it. The hag-fish is, on the other hand, caught abundantly by the liners when they shoot their lines too near a well-known 'hole' south-east of the Island of May, and the lamprey occasionally adheres to a fishing-boat. When the trawl is used on hard ground, wolf-fishes are often captured, but they are as common on board the liners on the same areas. Similar remarks apply to the dragonet on soft ground, and perhaps it is more frequent in the trawl than on the lines.

If a trawl were to pass over masses of the spawn of the herring at the moment the young fishes were escaping, the ground-rope and other parts would certainly do serious injury—just as a sweep-net on the rivers would to the tender young salmon lying with the large yolk-sac amongst the gravel of the spawning bed. A wound of the yolk-sac alone is generally fatal. The extraordinary multitudes of tiny young herrings (of the thickness of thread) occasionally carpeting square miles of

[1] *Vide Ninth Annual Report Fishery Board for Scotland*, p. 331, and Dr Masterman's interesting account, *Ann. Nat. Hist.*, Sept. 1893.

Wing of otter-trawl at stern of ship. The chain and block passing obliquely from the right is for hoisting. The cod-end of the net, with a rubbing-piece, is tied to the wing (for drying) on the upper right.

To face p. 52]

the sea—in St Andrews Bay, for instance, in March, would indicate the propriety of protecting them from all interference on this head, especially as they are then frequently just above the bottom. The very young sand-eels in the same way are often in great numbers just over the bottom, and, as this fish is perhaps next to the herring, the most important food-supply for the more valuable food-fishes, it likewise should not be interfered with. The enormous numbers of sand-eels on certain fishing-banks is one of the features of spring, and they are often mistaken by the fishermen for young herrings. The habit of the adult in burying itself in moist sand at and near low water is another illustration of the fact that a trawl may pass frequently over the same area, and yet leave, irrespective of those that escape through the meshes, a considerable number of animals untouched.

In the course of the ordinary trawling on the fishing-banks, no indication of the great number and variety of the post-larval and very young fishes is obtained. They are swept through the meshes, or otherwise escape. Yet, on sinking the great triangular mid-water net to within a fathom or two of the bottom, or by using the bottom trawl-like tow-net, crowds of the post-larval and very young food-fishes—flat and round—are captured; for example, those of the long-rough dab, lemon-dab, top-knot, ling, frog-fish, gurnard, dragonet, herring, and other forms. Neither trawler nor liner would seem to interfere at the stages indicated with these on such areas. In the same way, on other grounds, vast numbers of the young sand-eels, from 15 to 40 mm. (that is, from $\frac{5}{8}$ to $1\frac{9}{16}$ inches long) occur in the surface-nets in May, of which the only knowledge hitherto obtained was from the mouths of the fishes captured by liner or trawler, as neither interfered with them directly or indirectly in the pursuit of his calling.

The young fishes, on escaping from the pelagic eggs, are at first helplessly borne hither and thither by the currents, but the nature of their pigment, or even their transparency in most cases, probably proves protective. They are incapable of seeking shelter anywhere for a considerable time. When the flat fishes descend to the bottom their protective pigment on the upper side so closely mimics the sand or mud which they frequent that they are independent of growths, such as zoophytes or sea-weeds[1]. The very young round fishes again keep to the open water until they are of some size, and do not seem to seek either food or shelter at the bottom, and thus do not come in the way of the trawl. As they get older, many, such as the cod, coal-fish, and pollack, approach the rocky margins, swimming actively amidst the forests of tangles and other sea-weeds, and feed on the abundant animal life in that region. There they are also freed, for the most part, from interference by man.

The Recommendations of the Royal Commission on Trawling (1883—85) and the Method adopted by the Fishery Board for Scotland in carrying them out. Closure of areas.

It is now necessary to glance at some of the efforts of man to interfere with the marine food-fishes. As many are aware certain areas of the east coast have been closed for fully 12 years, and the whole of the three-mile limit for a shorter period. The former areas were selected in 1884 and 1885—very much at the author's suggestion—by the Royal Commission on Trawling and its lamented chairman, the late Lord Dalhousie, one of the most deservedly popular administrators Scotland has ever had. No one has watched with keener interest, therefore, the result of this experiment than the author, who felt a certain amount of responsibility, at least to science, on these important investigations[2]. After operations of more than

[1] These young flat fishes swarm all round our shores in April and May, on sandy and muddy flats, shallow rock-pools, estuaries, such as the Eden, and in the Swilcan burn as it debouches on the sands at St Andrews. They also, however, occur in the deeper waters.

[2] Those who read the Fishery Boards reports of to-day would imagine that the whole Trawling experiments and their adjuncts were the creations of the

twelve years' duration can it be said to-day that the closure of the areas of the Forth and St Andrews Bay or any other area has been followed by a notable increase in the numbers and in the size of the food-fishes therein? What evidence have the experimenters produced in regard to the oft-repeated statement that trawling on the same line soon exhausted the stock of fishes? What has been done—in the opportunity afforded by closure—to increase the valuable forms such as the sole and turbot in Scottish waters?

Before answering these, it may conduce to perspicuity if the various steps of the executive are briefly narrated. The areas carefully selected for closure—in 1884 and 1885—were, from south to north, the Forth, St Andrews Bay, and Aberdeen Bay. The Forth gives the results of closure in a great estuary or inland gulf having a considerable depth of water and various intrinsic sources of food-supply. St Andrews Bay is a typical shallow sandy bay on the east coast, with the estuaries of the Eden and the Tay opening into it. This area is more open to the North Sea than the Forth, and thus the "pulse" of the North Sea is more readily felt. The third, Aberdeen Bay, is wholly different from the others, since its perfectly open sea-board is defined—to the three-mile limit—only by artificial lines running eastward from Girdleness on the one hand and the Cruden Scars on the other, a distance of about 18 miles. It appeared to be important to have an accurate record of the seasonal variations of the food-fishes in Aberdeen Bay—for contrast and comparison with the two former areas, since the coast is even more exposed to the North Sea than St Andrews Bay.

The Reporters in the Sixth Annual Report of the Fishery Board (Prof. Ewart and the late Sir Jas. Gibson Maitland and their able assistants, the late Mr Jas. Duncan Matthews and Dr Fulton) fully understood one main function that was to be performed by the trawling experiments of the 'Garland,' viz. "what effect this mode of fishing was likely to have upon the ultimate productiveness of the waters around the coasts of Scotland, and especially in the territorial waters."

Board itself, whereas they were carefully discussed and mapped out by the Trawling Commission and especially looked after by Lord Dalhousie.

After a very brief experience, however, the Fishery Board ceased to examine the last area on the ground that "it has been impossible to obtain accurate statistics owing to the limited nature of the area closed, and for the reason that the takes of the 'Garland' have been extremely variable, and the results have not been considered of sufficient importance to warrant Aberdeen Bay being further closed for experimental purposes." There can be little doubt that the cessation of experiments in Aberdeen Bay arose from a misapprehension.

Carefully considered advice had been given to the Fishery Board by Lord Dalhousie, and he had exerted himself to procure the necessary supplies for carrying out the experiments in the selected areas—which he, as Secretary for Scotland, by and by closed. These experiments were to form the basis for future fishery legislation, and, therefore, their importance was beyond question.

For the carrying out of the experiments, the Fishery Board, possibly from lack of funds, deviated from the fundamental advice given, viz. that a powerful ship (like a first-class trawler) capable of carrying a 50-foot beam, and fit to proceed to sea in rough weather, should be procured, together with a smaller steam-vessel or tender for inshore work. The Board unfortunately selected an inefficient ship—both as regards the eastern seas and the manipulation of a trawl of sufficient size to search the grounds in an effectual manner. This has more or less crippled the experiments, especially as the work has only been conducted by day, and as a rule by a statistical, not a fishery expert, on board.

In carrying out the experiments lines were selected in the various closed areas, and along these the 'Garland' worked at intervals. Unfortunately, no regularity as to date, so important in making comparisons year by year, was maintained. This trawling along lines was the advice of the Trawling Commission, but the selection of them in the particular areas was left to the Board. At this moment the author is inclined to think that it would have been as well to make loops in stated parts of the areas so as to allow greater freedom in connection with wind and tide. Indeed, it may be a question whether perfect freedom should not have been given, so that the ship might, at uniform intervals, search the area in its most productive parts.

Shortly after the commencement of the operations the Fishery Board, as just mentioned, released Aberdeen Bay, and after a brief inquiry by aid of the 'Garland' in August, 1886, resolved to close the three-mile limit in the Moray Firth from the Ord of Caithness to Brora, Tarbat Ness, Balintore, then a point opposite the mouth of the Findhorn river, and thence to Kinnaird Head. This was done, it was alleged, in order "that valuable scientific results might be obtained were beam-trawling restricted in that district." In addition, the areas of the Forth and St Andrews Bay were increased seawards. The Board considered that the closure of the Forth and St Andrews Bay had already made signs of improvement both in the number and size of the less migratory flat fishes, a conclusion, on their own showing, subsequently abandoned.

The high average of fishes caught by the 'Garland' in the closed waters in 1887 still further seems to have influenced the Fishery Board in closing additional areas. Thus the territorial waters between Red Head and Kinnaird Head were closed in February, 1889, and thereafter the area of the Clyde. The 'Herring Fishery' Act of the same year, again, brought about the closure of the whole of the Scottish waters within the three-mile limit; and soon (1891) the entire Moray Firth from Duncansbay Head to Rattray Head was closed "to protect the fishes on their spawning-grounds (*e.g.* Smith Bank) and to ascertain the extent to which such measures are likely to be beneficial to the fish-supply."

The scientific grounds on which such action was taken have in the first place to be dealt with. The political or social reasons which may have had influence in effecting the closure do not at present concern us.

Now it will, in the first place, be apparent that before any reliable results could have been obtained from the experiments Lord Dalhousie's Commission mapped out, fresh areas were closed. The principle on which the closure was originally applied was that systematic and careful examination of these areas should be carried out for a considerable time. But such was impossible in the circumstances—especially of time and ship. Further, when the high averages of the 'Garland's' captures in 1887 in the Forth and St Andrews Bay are critically examined it is found that these were largely due to the fact that work was carried on solely during productive months, viz., May, June, August, and September, no winter month with its small captures reducing the average. The same misapprehension occurred later, when contrasting the captures of the 'Garland' during the first five years with those of the last five years of the period. The Fishery Board came to the conclusion that, since the captures during the first period were greater than in the second, deterioration had taken place in the protected waters—a very different finding from that (based on their earlier experiments) which they had formerly given as a reason for extending the closed areas. The Board now asserted "that there has been a diminution of the more important flat fishes in the closed waters, instead of an increase as was anticipated; and that this may probably be traced to the influence of beam-trawling in the open waters where the fishes spawn." Consequently it was urged that the Board should experiment by closing certain spawning-areas, such as the Moray Firth, and it was done.

No one will question the right of the Government to take such a step in the interests of the fishing-population—on philanthropic, social, or even on political grounds; but if such a step were taken on the basis of the scientific evidence furnished by the Fishery Board, then it would appear that the premises (and here all matters of fact are included) do not warrant the conclusion. There is, for instance, a serious misapprehension in contrasting an experimental period of five years in which a preponderance of work falls on the summer or productive months with an equal period in which the preponderance falls on winter or comparatively unproductive months. Let one instance suffice. In St Andrews Bay, in the first period, no hauls of the trawl were made in the unproductive month of February, whereas no less than 21 hauls occurred during this month in the second period. The two cases are thus widely different.

Changes in the Trawling-Vessels and their Apparatus[1].

With the exception of a few small sailing vessels and boats, trawling in Scottish waters is carried on, as it was in 1884, almost exclusively by steam-vessels; but, whereas at the latter period many of the vessels were old tugs or modified paddle-steamers formerly used for other purposes, most of the modern vessels, *e.g.*, those sailing from Granton and Aberdeen, are specially built for the purpose. The finest vessels do not cost much more than the serviceable vessels of the General Steam Fishing Company did in 1884, viz., £4500, but very considerable improvements have occurred in the arrangement and equipment. Some of these iron ships are 100 to 120 feet between the perpendiculars, and considerably more on deck, with a depth of 10 to 12 feet. The paddle-ships at Montrose[2] are 116 feet between the perpendiculars, 21 feet broad, and 10 feet deep; while the fine screw vessel is no less than 120 feet between the perpendiculars, 21 feet broad, and 11 feet 6 inches deep. The three latter have comparatively low bows, like many of the ships from Granton. The newer ships at Granton have also increased in size. Moreover, greatly increased height is given to the bow of the vessels at Aberdeen, so that the foothold on the fore-deck must be very uncertain, especially if slippery; but the water is kept out of the ship by such an arrangement. The after-part of the ship, however, is more or less flat, so that the trawls can easily be worked. These vessels range from 140 to 180 tons burthen, with engines from 40 to 65 horse-power.

Instead of having the steam-winch near the fore-cabin, in the newest ships it is placed on deck close to the engine-room, so that the steadiness of the ship is increased, and the bow kept out of the water. The screw-vessel at Montrose has two

[1] This account may be contrasted with that in the Trawling Report of 1884.

[2] Messrs Joseph Johnston & Sons.

winches, one being behind the foremast, the other (smaller) behind the mainmast. The latter is very useful in discharging fishes, and in working the dandy. In general, the Granton ships have the steam-winch in front, with the capstan behind, just before the engine-room, a different arrangement from that at Aberdeen. Moreover, a decided improvement is introduced by the presence of a 'brake' in connection with this apparatus. In 1884 reliance was placed on the old hawser fixed to the trawl-warp in the case of the net being held by a sunken wreck or a rock. Now, the moment the net is fixed, the 'brake' (which is secured to a moderate degree) permits the trawl-warp to run out, and thus save the net from serious rupture or total destruction while the ship is being stopped. In the Granton ships an iron-wire rope is used instead of a hawser from a hook on the mainmast, to save rubbing on the rail. This is fixed to the trawl-warp by spun-yarn. The length of the trawl-warp, which is of steel-wire rope, ranges from 200 to 240 fathoms. The warp has six outer and a central strand. The older warps had a Manilla centre, but the newer have wire. A change has also been made in the teeth of the wheels of the winch, for, instead of being transverse, they are now helical or oblique, in such ships as the 'Belcher.' The warp is run round a capstan in rear, and out by a slit with rollers in the bulwarks of the ship. The large ends of the winch are used, as formerly, for winding the bridles and all ropes and tackle, the latter being still the method of hoisting on board the bag of the trawl. Instead, however, of the snatch-blocks being fixed to the deck, they now are attached to the top of the engine-room. A considerable number of the paddle-ships still use a 9-inch Manilla hawser as trawl-warp, and it is wound round a capstan from wheels beneath the deck. These also have the piece of old hawser (at Montrose of about 13 fathoms) as a guard during trawling, but, as indicated, the best screw-trawlers have the 'brake' on the winch. In one or two of the older trawlers at Granton, the narrowness of the ship has caused the winch to be placed on the fore-part longitudinally, not transversely.

In the trawling ships of to-day the front wing of the otter-trawl is set free and the sole-rope and net follow it. The mate at the stern-wing, which has been made ready, waits till the other (wing) is at right angles to the ship and then lets go, the ship still moving, and the wire ropes are then both sent out to the required length, *e.g.* 150 fathoms on 50 fathoms' ground. When the ropes are fully payed out, the outer or bow-rope is hooked by a grapnel, and entered in a snatch-block at the stern-taffrail, so that both are near each other.

In some of the ships at Aberdeen the steering- or wheel-house has a roof, with side-panels and panes, so as to protect the men, and it occupies the same position, viz., in the centre of the vessel. Others have simply a canvas shelter above the wood. In one of the newest vessels at Aberdeen, the steering-house is open, as it is stated the men are apt to sleep in the covered houses, and prefer to be in the open air during their watches, while it is interesting to note that the Granton General Steam-Fishing Company's ships have always had open wooden wheel-houses. Besides a spirit-compass on a stand, a new vessel has an inverted one on a wooden pole, so that two are available in steering. Coals are still carried in the side-bunkers, which in the best ships have a floor of cement, so as to minimise the danger from fire. At Aberdeen small English coals are largely used[1], and instead of being piled loosely on deck at starting, as in some of the vessels from Granton, the extra coals are stored in bags, and are thus more easily handled. The finest vessels carry about 60 tons of coals in the side-bunkers adjoining the engine-room, and burn about 2½ tons per diem, with surface-condensing boilers; but fairly good ships often exceed this quantity[2]. The consumption of coal in such cases is, of course, a vital point in the economy of the trade, and a vessel which will consume 60 tons in 12 days is seriously handicapped. Some think that the larger vessels, which require more coal, are less fitted for remunerative work, since they catch no greater number of fishes. They might, however, be safer at sea. An improvement is the placing of the iron water-tank, which will hold about 270 gallons, under the deck,

[1] At 11*s*. per ton.

[2] This is much less than the quantity consumed by some of the old paddle-ships in 1884, *e.g.*, about 35 tons a week.

thus economising space and avoiding accidents. It is filled by a hose-pipe fixed to a screw-hole on deck. In the large screw-vessel from Montrose, the tank is placed behind the bulk-head of the engine-room, and a hand-pump raises water to the deck. In the newest ships an oil-tank, to hold from 40 to 50 gallons, is also filled beneath the deck in the same way.

The bulwarks of the new ships have self-acting scuppers for heavy seas, besides the usual small permanent ones, but no cement-gutters are now present at the sides, as it was found that they were rather a disadvantage, for, in such as have seen service, the cement becomes dilapidated. In 1884 the ships working off Aberdeen usually carried their fishes in covered compartments at the bulwarks in front, or they were even permitted to lie loosely on deck. This arrangement is now seldom seen, probably owing to the use of ice and the greater distances traversed. The bag of the trawl containing the fishes is emptied in the Aberdeen ships in a series of pounds (about five in number), formed by passing stout planks into upright grooves on deck in front of the winch, and in these the fishes are sorted and 'gutted,' preparatory to being placed in the fish-hold in ice. The labour involved by this method is a contrast to that of previous years off the eastern Scottish shores. Hence, when the catch at night includes haddocks of from 8 to 10 inches in length, these are considered unremunerative to treat in this way, and are thrown overboard.

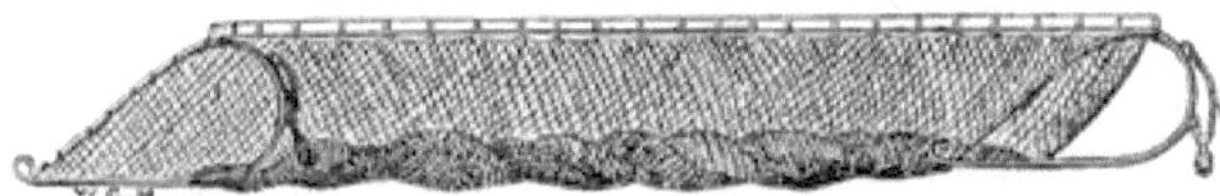

Outline of old trawl used for many years by the liners of St Andrews. The beam measured 35 feet.

In the ships of 1884 the stout boat was either carried on deck, or suspended from davits at the sides. It can be easily launched from the latter, but may be carried away, and, besides, the top-weight of the vessel is increased. At Granton the vessels formerly described have placed their boats on strong iron rails, 6 feet 6 inches high, on the starboard side, and bolted to the engine-room on one hand and the bulwarks on the other. A larger boat in the newer vessels is placed on rests in the centre of the ship over the engine-room, while in the most recent it occupies the centre of the stern, and the front 'stock' or support has a swivel. Moreover, in the 'Belcher,' the hook of the chain-lashing is jointed and fastened with a ring, so that the boat can be made ready in a minute. The modern boat is considerably larger, and is covered with canvas.

In connection with the fittings on deck, the use of raised or projecting figures or letters of sheet-iron on the funnel is one of the modern changes; they are very easily seen at a distance. The initial letter of the owner is sometimes added. Each vessel is, of course, marked likewise on quarter and bow.

The ice-house, which had just been introduced in 1884, is now an important part of the vessel, usually in front of the fish-hold. Five tons of broken ice are taken in the larger vessels to the distant grounds. It is sent from the stores in barrels, and passed from the cart to the hold by a funnel. So important has this feature become both for liners and trawlers, in Aberdeen for instance, that special factories have been erected for the manufacture of ice by the ammonia-system, about 90 tons being made daily in one[1] near the harbour, and 55 in another[2]. On the distant grounds, where most of the work of the larger vessels occurs, the ice is placed over the fishes after they are 'gutted' and consigned to the hold, as was done by the English trawlers from the distant grounds in 1884. The price of ice (at present 17*s.* 6*d.* per ton crushed and delivered alongside the ships) is thus an item of moment in the trawling expenditure. On discharging the fishes from Iceland, Faröe, or the Great Fisher Bank, the old ice is thrown overboard, and, though it might seem economical to keep it for use in a subsequent voyage, *e.g.*, for the preservation of the offal, for which 10*s.* per ton is got from the manure-companies, yet it is certainly the safer method. No wind-sails are now employed.

[1] North Brit. Ice Co. (Mr Lang's).

[2] Aberdeen Ice Co.

The fish-hold in the best ships is from 9 to 10 feet in height, divided into compartments, each with two shelves. In the 'Southesk,' a screw-vessel at Montrose, there are two holds. When fishes are stored with alternate layers of ice, the front of the compartment is closed with planks, unpainted or coated green with enamel-paint, which is readily purified by washing. The shelves, again, in each division, are useful in diminishing compression. This alone is a marked change on the Granton trawling-vessels of 1884, for the newest ships then had only an ice-chamber surrounding a central compartment in which the fish-boxes were placed. The smacks from Grimsby and other parts in England, it is true, used ice in the manner now described in 1884 and previously, but it was comparatively rare in Scotland at that period. It is necessitated now by the lengthened voyages to the more distant grounds. During the voyage the water which collects from the fishes and the melted ice is carefully pumped out by a 'donkey' engine, so as to keep the fish-hold dry. The hold will contain about 700 boxes of fishes, and great care is taken to keep it pure. In the Granton General Steam Fishing Company's ships ice is not used during the winter, for the fishes can be carried fresh to the market by means of one ship acting as a 'carrier' daily. In the warmer weather, however, ice in bags is taken on board each vessel. Few ships at Granton, indeed, have the compartments for packing the fishes in ice, with the slips of board for closing them. This shows that the majority fish in the less distant waters.

In some of the newest vessels the accommodation for all the crew is in the aft-cabin, the fore-part of the vessel being relegated to the fish-hold and stores. This appears to be a decided improvement in regard to the maintenance of a cool temperature and pure air near the fishes, especially when long voyages are undertaken. Formerly the crew had a fore-cabin, and the captain and mate an aft-cabin, and in many vessels the same arrangement still occurs.

The engine-room of the newer vessels is better ventilated, and the arrangements for the working of the engines facilitated. Even the ventilators are utilised for the hoisting of cinders from the hold by the aid of a small windlass. Moreover, in one the engine-room has an entrance from the galley as well as from the side, a convenience in stormy weather. A feature in contrasting the ships at Granton and Leith with those at Aberdeen is the small elevation of the engine-room above deck in the former.

In some ships the shrouds from the mizzen-mast are fastened to the deck about a yard from the bulwarks, so as to leave a clear space for working the trawl. In the larger ships, however, this is not necessary, the space in rear of the shrouds being sufficient for the trawl, or shrouds are altogether dispensed with, as in the Montrose paddle-ships, which have only a foremast.

The galley for the cook is in many under the bridge in front of engine-room, or in some in the forecastle peak; but in the Montrose paddle-ships it, with the water-tank, is at the side near the paddle. These, also, have two tow-rails, one in front of the cabin for the crew, and one behind the cabin for the captain (aft), as the vessels are used for towing. A hand-windlass for raising the anchor is also present.

The average length of the trawl-beam in the best ships[1] is 54 feet, it being found that a longer beam does not work so satisfactorily or catch so many fishes. At Montrose the beam is 52 feet. As before, it is composed of two or three pieces of oak or French elm, though occasionally it is in a single piece, and has a diameter varying from 10 inches to a foot. The shape of the iron trawl-head is scarcely altered, the posterior iron plate in a few being somewhat more abrupt than in 1884, thus conforming to the English type of trawl. The height of the beam from the ground varies from 3 feet 8 inches to about 4 feet. The 'Athole,' one of the General Steam Fishing Company's ships, was in 1894 provided with an 'otter' trawl[2] with large wooden ends about 12 feet long by 5 feet broad, which take the place of the 'hammer' of the pole-trawl described by the Commissioners of 1863[3], and which are simply

[1] In 1894.

[2] The invention of Mr Scott.

[3] *Report, Sea Fisheries of the United Kingdom*, vol. i., Appendix, p. 3.

the much-enlarged wooden ends in use in the otter-trawls in the Forth in 1858. These huge wooden (door-like) ends have on one side in front two powerful iron bars meeting to form a V, and supported by two accessory stays, the whole forming a projecting apparatus to which the chain connected with the warp is fixed. Towards the rear a perforated iron plate gives passage to two chains (one from each of the powerful iron bars above-mentioned) for the attachment of the swivel for the trawl-net. The lower edge of the wooden end is weighted anteriorly with a heavy bar of iron, which occupies nearly half the length of the apparatus. A special and powerful rectangular frame of wood, with a top snatch-block, is fixed at the port-bow and taffrail for hoisting the ends on board; and they form a striking feature from a distance, as—with the boards—they project 6 or 7 feet above the bulwarks. The foregoing trawl is said to capture cod more freely than the beam-trawl, as many as 20 score having been secured in one haul. It has been adopted by most of the modern trawlers, a much greater width of net and increased powers of capture being possible with this apparatus.

It was formerly pointed out that, when the iron trawl-head was dislodged, great difficulty was experienced in repairing it—especially in rough weather. The new trawls at Aberdeen had in 1894 a broad band of iron which bends round the end of the beam, and on which the loop of the trawl-head goes. It is secured by an iron pin and safe. This sheath protects the end of the beam, and must save much time at sea. At Granton the ends of the trawls were guarded by flat iron plates, but they did not form a loop over the ends. The trawl-heads were secured by a pin, as already mentioned. In the finest ships the length of the trawl-net is about 118 feet, and the arrangement is as follows:—For the first 56 feet next the beam the mesh of the net is 3 inches from knot to knot; the next 38 feet has at first a 2½-inch mesh, diminishing to 2 inches towards the posterior end, while for 24 feet the bag or 'cod'-end of the trawl-net has 1½-inch mesh. At Montrose the trawl-net consists of 44 feet of 3-inch mesh next the beam, then 44 feet ranging from 2½ inches downwards, while the last 14 feet has 1½-inch mesh. There is thus no diminution of the mesh at the 'cod'-end. Moreover, no improvement in the shape of a 'bonnet' or apparatus for preventing the compression of the fishes has been found serviceable notwithstanding various recent statements to the contrary. The net has various rubbing-pieces of old net and 'bass' ropes, and the usual pockets internally. The ground-rope is variable in composition. The majority have this part of the trawl composed of rope only—an outer layer being wound round a central rope. The ground-rope of the Montrose ships is of Manilla soaked in tar, 8 inches in circumference, and made up to 13 with others twisted round; and in the finest ships elsewhere it is 124 feet long. In some, two pieces of chain are inserted at the ends, thus making three divisions of the ground-rope, viz., a central, entirely of rope, and two lateral, with a centre of 18 feet of chain, each being tied to the other with spun-yarn. Ground-ropes with chain throughout are not now used. In certain ships the ground-rope has a centre of wire-rope with a series of wooden rollers, with occasionally, here and there, a pair of metal rollers (12 in all—Gunn's patent). The rope is also in three divisions, and costs about £6. 10*s.*, or 30*s.* more than the ordinary form composed only of Manilla ropes. This arrangement is thought at Aberdeen to give an increased catch of fishes—sometimes about 5 or 6 baskets more than by the ordinary ground-rope. In some ships, again, the port- and starboard-trawls have each a different ground-rope; in the one the rope is all of one piece, whereas in the other three breaks occur, viz., two of wire and a third of chain. In one ship, the ground-rope had only 8 feet of chain at each end, while the centre had rope. All, however, do not think that the rollers are so satisfactory as a ground-rope with pieces of lead in the centre. Moreover, one of the features which contrasts strongly with the condition in 1884 is the fact that the newest ships, with the exception of the Montrose vessels, now carry two trawls—a starboard- and port-trawl—complete in all respects. This arrangement has been in force for at least four or five years, and probably was introduced from England. At Aberdeen, however, the second trawl is, as a rule, used as a reserve-apparatus, and is not put into requisition until the first has received damage. The mode of working the two trawls would thus appear to differ materially in the respective countries, since, according to an interesting paper by Mr W. L. Calderwood[1], as soon as the contents of the first trawl are placed on the deck, the second is immediately 'shot' overboard. The same arrangement has been found at Grimsby by Mr Holt, who mentions, however, that the reserve-trawl is shot while the 'cod'-end with its fishes is still hanging from the tackle. The General Steam Company's ships at Granton (nine in number) have not varied in regard to the single trawl-beam, but they carry a second net. Consequently, the large snatch-block and rollers occur on the port-side only. As before, the net is attached to the trawl-beam by grummet-lashings or by cord. The other parts, comprising the dandy and bridles (each about 25 fathoms) and the chain for the former do not differ materially from previous descriptions. The steel-wire rope is about the same length, viz., 200 to 240 fathoms, and lasts about ten months. The aluminium trawl-warp does not seem to have met with favour in Scotland. In some ships it is not, as formerly, left on deck after the check of wire-rope is fixed to the mizzen-mast, but carried outside the bulwarks, so as to avoid accidents to the men. Those which, like the Montrose ships, use a Manilla rope (generally about 180 fathoms), require a new one every six months, the old one being utilised in preparing ground-ropes.

The shooting of the trawl is carried out in a similar manner to that of 1884, only there are no trawl-davits at the taffrail in the best Aberdeen ships; and, instead of the snatch-block then in general use, more convenient 'dandy' scores (snatch or tumbling blocks), of which Sudron's or Scisson's patent are the best. At Granton and Montrose the trawl-davits are still in use, with snatch-blocks on deck. The lid of the block is opened during trawling. The trawl-warp leaves the drum, passes round a capstan, and out through rollers, either on the port- or starboard-side, according to the trawl in use. Blocks on the mizzen-mast are still employed to hoist the

[1] 'British Sea Fisheries and Fishing Areas,' *Scottish Geogr. Mag.*, Feb. 1894 p. 73.

stern-end of the trawl, and the fore-mast has a derrick. In 'shooting' the trawl the ship goes at full speed. When the 'cod'-end of the trawl is unshipped, the mate at the same time orders the fore-trawl-beam lashings to be freed, and when the beam is at right angles to the ship the 'stopper'-rope is let go, and the order 'ware forward' then sends off the trawl-warp from the drum.

A better arrangement now exists for assisting in unshipping the heavy trawl-heads, for these rest on a stout wooden platform about 18 inches high, and thus are easily swung over the rail; and, besides, the deck is saved from injury. In one or two ships at Granton larger platforms for the fore-end of the trawl have been fitted. In a new vessel, indeed, a square of plate-iron has been put on the deck at the point most injured by the trawl-head. In rough weather a chain fastens the trawl-head to the nearest iron stanchion at the bulwarks, and is used in bringing the front trawl-head on board. In the same way an additional chain at the stern-end is sometimes useful. In the Montrose paddle-ships the wheels for winding the trawl-warp (a Manilla rope) are below, and only the capstan is on deck. The latter (capstan) in some trawlers is made too high, and is wrenched out of its fastenings.

The trawl is usually down for five hours on the 'Great Fisher Bank' and other grounds, though trawlers working near home regulate the time rather by the nature of the bottom than anything else, in some cases spending as much time (three hours) in mending the net as in trawling on hard ground, or where wrecks and anchors occur. The trawling-period, indeed, on hard ground is about three hours, on soft ground five hours. When productive ground is discovered, a 'dan,' or buoy, with a red or black flag by day, and a white globe-light, close to the surface, at night, is put in the water to mark the spot, though it is liable to be carried away by other ships, and the lamp broken. This buoy has a pole, with heavy iron bars, at one end, and towards the other about ten flat pieces of cork, upwards of a foot square. In one or two ships floats of skin—such as the liners use in herring-fishing, with pole and flag, were substituted for the cork buoys, or small pieces of cork on a string. The rate of speed when trawling is, as formerly, about 2½ knots an hour, though on muddy ground a higher rate is sometimes maintained. In sailing, the best ships go about 11 knots. At night the captain and mate take watch alternately—with one of the crew.

The crews on board the trawling ships remain very much as in 1884, the usual number being eight, though there are only seven in the Montrose paddle-ships, one of whom is cook. The latter may be either an old man or an adolescent. Each is furnished in the newest ships with a life-jacket of cork, and there are besides two life-buoys on deck. Only two at Aberdeen, the captain and mate, now have a percentage on the amount of fishes captured. The rest of the crew have ordinary wages. At Montrose the captain and two fishermen have a share in the 'catch'; the rest have wages. There are seven men on board the ships of the General Steam Fishing Company at Granton, instead of eight as formerly. The percentages given to each remain almost as in 1884, a graduated series running from the 'deck-hands' to the captain. The first engineer gets 5*s.*, and the second 3*s.* 4*d.* per ton of fishes.

In 1884 the Granton General Fishing Company's ships used 'cringles' in transferring, during stormy weather, the fish-boxes to the 'carrier' for the day. This practice has now been abandoned, and the ships either run to quiet water, and place the boxes on the deck of the 'carrier,' or they are at once transferred by boarding. It is during the latter operation that considerable injuries occur to the bulwarks and rail of the ships, the former having the stays bent, and the latter being frequently driven in.

One of the newest ships[1] at Aberdeen is a steel vessel—with a well—for fishing at Iceland and Faröe. It is 103 feet between the perpendiculars, and 114 feet on deck, 21 feet broad, and 12½ deep. The well is one of Houston and Mackie's patent fish-wells, and occupies the entire centre of the ship, the roof of the well sloping inward about half-way up the side of the ship, and leading to the hatches—the opening thus being much smaller than the bottom. The water accordingly will be some-

[1] 'Ocean Bride'—Mr Drummond's.

what steadied during the motion of the ship, though, as the cod will have a roof to rub against as well as walls, injuries may readily occur. The water is driven in during the voyage, rises to the surface of the well, and overflows by an opening in the side of the ship. A constant current is thus kept up. A grating at one end (the lower) permits the removal of refuse from the bottom of the well. While the cod swim freely in the tank, the halibut are tied, as usual, by the tail to the iron rail at the margin. The vessel has been specially fitted for the capture of these by hook and line; and at present no trawl is aboard, though such can be shipped at any time, and the newest apparatus (*e.g.*, steam-winch and Sudron's patent dandy-score) is in readiness. The foremast has a derrick-boom, and the anchor-winch is worked by steam. The boat rests on a swivel-stock on the port bow, and is intended to be used as an accessory well. The cabins for the crew (viz., captain, two engineers, and nine men) are at the stern, while in the high bow is a store, and behind a convenient hold for fixing the bait (herring). An ice-house, fish-hold, and all the newest fittings in the engine-room and other parts show the care that has been bestowed on the construction of the vessel. The consumption of coal is estimated at 3 tons daily.

Ships similar to the foregoing have been employed for some years at Grimsby for line-fishing in Iceland, but several improvements have been introduced in the new ship. Moreover, when required it can also be used as a trawler.

Recently a new form of trawling-apparatus has been introduced chiefly by the line-boats though likewise by a few trawlers. This consists, in the line-boats of a seine-net of about 160—180 yards, and about 6 yards deep, in the Clyde 5 yards deep, made of stout cord (No. 14 twine), the mesh being about 8 inches long,—having a rope at the top with corks and a rope at the bottom, pieces of leaden pipe being placed on the latter at intervals of 6 feet. The centre of the net has a bag about 20 feet long, and about 12 feet at the aperture, the mesh of which is smaller—with a 'cod'-end for opening. A float is attached to one end, and the net is thrown out in a straight line for nearly three-fourths of the length, when the boat makes a circle to join the float. Both ends of the net are hauled, and the contents are carried towards the central bag. The net sweeps the ground, and fair catches of flat fishes, especially plaice, turbot and skate, have been made.

In the steam vessels the net is much longer, probably about 800 yards. Great labour is entailed, since the net must frequently be hauled.

St Andrews Harbour in its Trawling-Days.

Changes in the Line-Boats and their Apparatus.

The changes which have taken place in line-fishing have been more slowly developed, and have been modifications of the very ancient system which was prevalent in pre-christian times, and has been handed down from generation to generation. Any fishing centre will suffice for illustration, and we shall take St Andrews. Up to about the year 1860, the largest boat employed was about 35 feet long, clinker-built, with pump of wood, and without a deck, and fishing was carried on by the crew of six men not far from land. The lines consisted of three strings and a half or 700 hooks (each hundred of the fishermen, however, being equal to 120). The mainmast carried the large lug-sail, and the foremast the smaller. Instead of a fixed stove for cooking the men used an old pot, or a small portable fire on an old tray, but as they did not go to distant grounds this temporary arrangement sufficed. Their fishes were placed in baskets—each containing about ten stones. No market existed in those days for fishes, and they were sold by the women in town and country, or to "fish-cadgers," who carried them in carts to neighbouring towns. Their captures were accordingly limited in quantity, and a large catch could not easily be disposed of. Towards the beginning of the century, indeed, a cart has been filled frequently for a shilling with fishes of various kinds.

For herring-fishing the nets were of hemp, had 29 rows to the yard, and did not shrink. Each boat had eighteen or twenty nets, each being 55 yards in length and 15 yards deep. The nets were hauled by the hand, as in primitive times. In white fishing a gaff or "clip" was used for seizing heavy fishes or those unhooked when hauling the lines.

Now the position is much altered. After the period mentioned the length of the boats increased, 4 strings instead of 3 formed the lines, and now no less than 6 strings with about 1,200 hooks are used, while the hooks instead of being 36 inches are 45 inches apart. The largest boats are 60 feet or a little more in length, are decked, and furnished with an aft-cabin and berths for the men, as well as a cooking-stove. They are carvel-built (that is, instead of the planks overlapping as in the clinker-built boats, the edges touch, and the intermediate chink is caulked). The foremast is the larger and carries the heavier sail The pumps are now of iron. The greatly improved fittings of the boat, such as patent blocks and the arrangement of the ropes, enable a smaller crew to do the work, so that occasionally 5 men suffice in a boat of 60 feet, whereas in the olden time there were 6 in the 35-foot boat. The nets, of which each boat has 100, are 60 yards long, and 18 yards deep, and they are made of cotton, with 33 rows to a yard, and after two years' wear, the shrinkage gives 36 in a yard. Instead of manual hauling, there was first the improvement of a net hauler (see Fig. adjoining) and lastly the same worked by steam. The gradually increasing size of the boat rendered long hand-nets necessary for capturing stray fishes over the gunwale. Instead of baskets, boxes have been used for the fishes for nearly a quarter of a century, the box, indeed, making its appearance with the trawls and increased facility in disposing of the fishes. Finally, in the most advanced towns steam-liners have largely increased.

It is interesting to notice the parallelism in the increase of the nets and hooks, for while no one brings forward the enormous increase of the nets as proof of the decadence of the herring, it is otherwise with the increase of hooks, which, for instance, in the Reports of the Fishery Board, is held to prove the diminution of white fishes in our seas, and the call for restrictive legislation. Changes have occurred in both groups (herrings and white fishes), but they are apparently not of a kind to be dealt with in the manner just mentioned.

THE PRESENT STATE OF THE LINE AND TRAWL FISHERIES IN RELATION TO THE FISHING-GROUNDS AND THE FISHES[1].

In 1884, under the head of 'General Remarks,' as careful a survey of the situation of the fisheries in connection with both line-fishing and trawl-fishing was drawn up[2]. In reading over these remarks at the present time the position does not seem to have been misunderstood; indeed, there is little at variance with the condition as now shown by fifteen years' experiments and observations. Amongst other remarks it is stated that 'steam-trawlers at present can only fish profitably within a

[1] *Vide*, the Author's '*Brief sketch of the Scottish Fisheries*, 1882—1892,' Dundee, 1892.

[2] *Vide Report of the Royal Commissioners*, pp. 377—380.

Latest hand-apparatus ("Champion") for the hauling of herring-nets. It is worked by two handles.

To face p. 74]

Baiting the line. The "shelled" mussels are on the board on the left.

To face p. 75]

moderate distance of the land; and were the fishes to become so thinned that, with all the skill and energy shown in managing the ships, the returns proved unsatisfactory, trawling might voluntarily disappear. There is no reliable evidence, however, that before such a result would happen irreparable injury would have been done to the sea-fisheries.'

Now, at that time there were in Scotland a total of 61[1] trawling-vessels—of which probably about one half were steamers, the other half being sailing-boats or vessels used for trawling. The exact numbers cannot be obtained, but there were from 12 to 20 boats used in trawling at St Andrews, 6 to 8 came from Broughty Ferry, 2 or 3 each from St Monans and Cellardyke, and others existed in the Moray Firth. Trawling, indeed, at St Andrews was an old custom, the Buckhaven fishermen having introduced it early in the century, and subsequently the local fishermen carried it on more or less regularly, generally trawling in September and October, and in March and April, though occasionally much longer. The frequent presence, however, just before the period of the Trawling Commission, of 10 or 12 powerful steam-trawlers to compete with them on their own ground quite altered the aspect of affairs. The energy with which the steam-trawlers generally worked—for trawling went on by night as well as by day, and in weather unsuitable for the liners—introduced in Scotland a new era into the department. Fishing was to be carried out no longer by more or less independent crews, bound together by blood-relationship or other ties, and whose working hours were largely regulated by the weather and tides, or their own convenience and necessities. Yet, their whole domestic life was interwoven with the time-honoured pursuit. Their wives and daughters laboriously baited the hooks and arranged the lines in the baskets for 'shooting,' they gathered the 'bent'-grass for separating the layers of the line, and with the sons dug lob-worms or procured mussels for bait. In the olden time, indeed, their wives and daughters were likewise their fish-merchants, and disposed of their captures to the best advantage.

[1] The numbers are taken from the *Report of the Select Committee of The House of Commons*, 1893, p. 396.

Now (1883) active and powerful vessels, propelled by steam, and thus more or less independent of the weather—manned by a captain responsible to owners or their manager, a crew bound together only by discipline and pay, and whose fishing apparatus required no bait, appeared on the field. Further, instead of following the pursuit on grounds familiar to generations before them, the new fishermen not only ranged over these, but sought new and sometimes more distant fields. Capitalists took up the question, and fitted out powerful ships in both Scotland and England, and sent them into Scottish waters, so that the liners met with most formidable rivals. The complaints of the line-fishermen at this period (1883) and subsequently necessarily attracted much attention, and great sympathy has always been expressed in regard to their condition, for undoubtedly the larger and more regular supply of fishes had a tendency to diminish value, and this caused a reduction of income to the liner, and the fishes on certain of the nearer grounds were thinned, and perhaps rendered more wary. In the Report of 1884 it was said that 'two competitors are in the field instead of one, and for the liner it may take closer work, even with all the help improved modern appliances in boats and material can give, to keep pace with his rival;' and further, that it would be a great calamity if any mishap should befall such a fine race of men—hardy, willing, and adventurous. Complete destruction, or, at any rate, most serious interference with the fishing-grounds, and the destitution of the fishing-population, were then predicted, and many anxious eyes watched the development of events, since about 45,000 men at least were dependent on the net- and line-boats of the country, whereas only a few hundred—perhaps between 200 and 300—were at that time engaged in the trawling industry.

Fifteen years have elapsed since this change took place, yet so far as can be ascertained, the condition of the sober and industrious liner of to-day is far from being destitute, and many are able to utilize the Savings Bank, to acquire property or a competency for their old age. It could not well be otherwise, when, with all the uncertainties of fishing, four men will occasionally have about £26 a week, or a single 'shot' which produces £9 or £4 in white-fishing. In herring-fishing, again, a crew of two boys and a man will make £9 in a single night in the Forth, and this in the year 1898. In most places they have enforced idleness from storms, and there is little for them to do on shore. No other class of working men, indeed, more frequently have such periods of rest, and the fishermen are thus conspicuous features in the every day life of their towns and villages. Some compensation is in this way given for their hazardous calling, and for their courageous battle with the elements, and, it cannot be doubted that it is the interest of the nation to encourage them even more than those who follow other trades.

The experienced remarks of the late Mr Spencer Walpole on the subject of the British fisherman's financial condition[1] is worthy of the earnest attention of all interested in the Fisheries, and more especially of that class whose persistent aim, unless we are to suppose utter incapacity, is less to benefit the Fisheries and fisherman than to make progress—political or otherwise—in a different direction. Every word of Mr Walpole's remarks applies equally to Scotland as to England, and in every essential particular can be supported from personal experience. He specially points out the errors into which Prof. Leone Levi fell when he supposed that the fisherman was not half so well off as the ordinary agricultural labourer. He truly says "I do not think that anyone who has any acquaintance with the fishing community will endorse that statement. If you examine an ordinary fisherman's dress, you will find it warmer and more costly than that of the labourer;...he consumes a larger proportion of animal food than the labourer;...you will find rows of cottages not merely occupied, but owned, by fishermen, built or purchased out of the profits of the fisheries....Many of them own their own nets and lines, and some of them have a share in the boats in which they sail....Many of the masters are boat-owners, with £250 to £1500 of capital—who have begun their lives as ordinary fishermen."

In communities of sober and industrious fishermen comfort is almost always present, and want unknown.

[1] *Official Rept. Internat. Fish. Exhib.* vol. xiii., pp. 173—178, 1884.

The crews of the average steam-trawler are not less daring than their comrades of the line-boats, but they lead a much more active life, and have little or no time, except perhaps in great storms or on Saturday and early on Sunday, for shore-life. Their constant services by night and by day throughout the week must frequently test even the hardiest constitution, which can only be kept in condition by a faculty of snatching a little repose when off duty during the time the trawl is down. Their average earnings are probably greater than those of the liner, but as a rule, they work much harder, and permit themselves little leisure. As their remuneration depends in most cases on the success of their fishing, it is no wonder that they select the richest grounds within a reasonable distance, and leave the less lucrative areas to others. It would be a great boon to these men if their Sunday on shore were unbroken.

Since 1884 the trawling vessels in Scotland have steadily increased in number, so that within ten years they have been considerably more than doubled: the returns for 1893 showing that there are no less than 142 vessels and 720 men thus employed,—the total value of vessels, exclusive of gear, being about £240,737. Or, to go more minutely into details, of this number 72 are steam-trawlers, having a tonnage of 2625 tons, and valued at £237,004, to which has to be added the fishing gear, £10,746,—making a total of £247,750. These vessels are manned by 544 men. The rest (70) are sailing-trawlers, having a tonnage of 423, and valued at £3733, while their gear is estimated at £1332—making a total of £5065, with 176 men on board.

In addition to the foregoing there were 39 steam-trawlers belonging to English owners, fishing regularly from Scottish ports, and the tonnage of which was 959 tons, value £92,100, and value of gear £3850,—making a total of £95,950. These had 296 men on board. The disproportion between the number of men employed and the cost of the material is chiefly brought out when it is mentioned that for 1892 the liners and net-fishermen were 45,629, while they had 13,865 boats, valued at £680,000. It will thus be seen that, while the average is

Ordinary Fishing-boat in Aberdeen Harbour, August, 1898.

To face p. 78]

about £1695 for each trawling ship, for the liner it is about £49. The disproportion, again, in the trawling vessels, between the first-class and the small sailing-boat, *e.g.* of the Clyde, is very great,—the former being about £5000, the latter under £40.

If the returns of round, flat, and other fishes landed, irrespective of herrings, sprats, sparlings, and mackerel, which do not prominently bear on the present question, be considered, it is found that in 1892[1] the liners brought to shore 1,229,809 cwts. of round fishes, viz., cod, ling, torsk, saithe, whiting, haddock, and conger, which realised, at 8*s.* per cwt., £516,524; the trawlers landed 261,200 cwts. at 10*s.* 11*d.*, or £143,062; the liners produced 100,228 cwts. of flat-fishes, viz., flounders, plaice, brill, skate, halibut, lemon-dabs, and turbot, at 10*s.* 9*d.* = £53,973; the trawlers, 77,649 cwts. of flat fishes at 25*s.* 4*d.* = £98,295; while of other kinds of fishes, which include hake, bream, gurnard, cat-fishes, and sillock, the liners had 61,224 cwts. at 4*s.* 9*d.* = £14,646; and the trawlers, 41,256 cwts. at 4*s.* 7*d.* = £9,410. The total in each case are, for the liners 1,391,261 cwts. and £585,143; for the trawlers 380,105 cwts. and £250,767.

In glancing at the returns (1892) of the Board, which were handed in by the late able Chairman (Mr Esslemont) to the Select Committee in 1893, it would seem that one fish, viz., the green cod or coal-fish, is included both amongst the round fishes and the 'other kinds of fish,' in the former having the name of 'saithe' (adult), and in the latter 'sillocks' (young), but this is not a point of much importance in regard to the results. As might be expected, the liners, and notably the long-liners, have the predominance in the round-fishes, especially in regard to cod, ling, and conger, the latter being apparently seldom caught in a trawl on the Eastern coast. These large fishes, moreover, would appear to protect themselves to a considerable extent from this apparatus, especially when it is in frequent use, so that it is only in water that is disturbed by gales or by working at night that they are caught in numbers

[1] The full value of the labours of the Royal Commission of 1883, and especially of the late Lord Dalhousie, in establishing a series of proper statistics for the fisheries of Scotland, cannot be over-estimated.

under these circumstances. Nor is this surprising, since even a tiny cod, of little more than ⅛th of an inch, can avoid the forceps intended to capture it. The cod and saithe are also largely caught by gill-nets on the West coast; while the great lines, carrying hooks baited with herring, are the chief means of capture used in the case of the conger. Further, it has to be remembered that the trawlers, both near and distant, as a rule, throw overboard their small haddocks (8 to 9 and 10 inches), in both cases because it is not worth their trouble to bring them to market and pay dues for the trifling sum obtained for them; and in the instance of the distant trawler, to avoid, in addition, the labour of 'gutting,' and the expense of ice. Yet the liners bring these to market and they are included in their returns. It is an interesting fact that, notwithstanding the recent remarks concerning the condition of the trawled fishes, that the price of the latter surpasses that of the former by 2*s.* 11*d.* per cwt. It is true the trawler can more readily reach the market with his fishes, but against this has to be placed the great number of local fishing-boats which have only brief distances to traverse, and the fact that the trawlers who go to distant banks bring fishes 'gutted' as well as preserved in ice, and the appearance of which is not always in their favour.

When the flat fishes are considered, it is found that though the liners produced considerably more in weight, yet the price obtained per cwt. is not half (by 3*s.* 10*d.* less) that got by the trawlers, so that the total value of the flat fishes procured by the latter is nearly double that of the former[1]. Yet we know that halibut are largely caught by the liners, and that the three-mile limit and the closed waters in addition are at the disposal of the latter for relays of lines wherewith to capture plaice, dabs, and flounders. In all probability, however, it is the plaice, the witches, and especially the lemon-dabs and the turbot which prove so advantageous to the trawlers.

Of the 'other kinds of fish,' little need be said except that

[1] At Montrose, for instance, the flat fishes landed by trawlers realised nearly 20*s.* per cwt., while those caught by line produced only 9*s.* 11*d.* per cwt. But turbot alone was sold at £3. 6*s.* 2*d.* per cwt., so that the trawlers had the advantage in this respect.

comparatively few hake come into the trawl, whereas the liner perhaps obtains a larger number; that while the liner brings the gurnards to shore and often eats them, they are frequently thrown overboard by the trawler; and that the cat-fish (wolf-fish) is caught by both in considerable numbers, but whereas, in certain trawlers, this fish is taken to port on Tuesdays, it is thrown overboard at the end of the week.

In 1893 the equivalent returns show that the liners brought to land 1,136,389 cwts. of round fishes = £466,399, this being 93,419 cwts. and £50,125 less than last year. The most marked deficiency was in haddocks, 69,766 cwts. and £35,092; cod, 54,260 cwts. and £20,661; and whiting, 15,381 and £5,741. An increase had taken place both in line- and trawl-fishing in the other round fishes, viz., ling, torsk, saithe, and in the conger caught by line. How far this diminution was due to the unfavourable weather of 1893 is an open question. It certainly must have had some influence. The abundance of very small haddocks is another fact to be remembered, since many were not brought to shore, and they occupied hooks on which larger fishes might have been caught. The trawlers landed 309,862 cwts. of round fishes = £178,304, or 48,662 cwts. = £35,242 more than last year. With regard to flat-fishes, the liners produced 57,149 cwts. = £43,306, or 4,685 cwts. and £48 more than last year; the greatest increase, 5,594 cwts. and £566, having been in halibut, but these apparently were largely caught in distant waters, such as off the coast of Norway, Iceland, Faröe, and elsewhere, so that they confuse the returns from British waters. An increase also exists in "soles" (lemon-dabs) of 50 cwts. and £120. A slight decrease, again, occurs in flounders, plaice, and brill. The trawlers landed of flat fishes 71,024 cwts. = £89,781, a decrease of 604 cwts. and £7,243 on last year, this decrease being largely due to the deficiency of lemon-dabs, viz., 6,133 cwts. and £8,448, and a deficiency in turbot of 94 cwts. and £762, while an increase occurred in halibut and a larger increase in flounders, plaice, and brill, 5,197 cwts. and £1,597. This year skate form a separate return, which shows that the liners produced 52,626 cwts. and £10,725, or 4,862 cwts. more than in 1892, yet with only a trifling excess of income over that year, viz., £9. 10*s.*, a result probably due to diminished prices. The trawlers landed 5,383 cwts. = £1,015, or 637 cwts. and £253 less than in 1892. Of 'other kinds of fishes' the net-fishermen brought 3,517 cwts. = £891, or 102 cwts. and £731 more than in 1892, while the liners landed 46,461 cwts. = £10,726, or 11,347 cwts. and £3,160 less than in 1892. The trawlers again caught 39,418 cwts. = £9,215, or 1,838 cwts. and £195 less than in 1892.

The price of the round fishes in 1893 is respectively for the liner 8*s.* 2½*d.* per cwt., and the trawler 11*s.* 6*d.*, or a balance of 3*s.* 3½*d.* in favour of the latter, and therefore a higher proportion than in 1892. In regard to flat fishes the inclusion of skate makes a considerable difference; thus the average price for flat fishes, inclusive of skate, is for the liner 9*s.* 10*d.*, for the trawler 23*s.* 9*d.* per cwt., whereas, when the skate are excluded, it is for the liner 15*s.* 2*d.*, for the trawler 25*s.* 3*d.* In the former case the trawler receives no less than 4*s.* 1*d.* more than double the amount obtained by the liner; in the latter case the trawler receives 10*s.* 1*d.* per cwt. more than the liner. The disproportion in any case is marked. In connection with prices, however, it has to be borne in mind that in many cases the liner is compelled to sell his fishes in remote districts or unfavourable markets, whereas the trawler takes care to put his fishes into the best market, and in quantity.

Again, the grand total of all kinds of fishes landed in 1892 was 5,436,138 cwts. If herrings, sprats, sparlings, and mackerel (viz., 3,664,771) are deducted, 1,771,367 cwts. are left, of which 1,391,262 cwts. were caught by liners, and 380,105 cwt. by trawlers, or, in other words, the liners caught more than three times the quantity of fishes landed by the trawlers. In 1893 the grand total of all kind of fishes notably exceeds that of 1892, and is no less than 6,208,018 cwts., or 771,880 cwts. more than in 1892. The greater proportion of this, however, is made up of herrings, viz., 4,486,187 cwts.,—that is to say, a fish which is more or less unprotected at all stages of its life is apparently able to hold its own against its destroyers. It is, however, a purely pelagic form, and depends on the pelagic or floating fauna for its food. If the herrings, &c., are deducted,

Repairing the lines.

To face p. 82]

a balance of 1,721,831 cwts. is left for the liners and trawlers, being 49,536 cwts. less than in 1892. Of this 1,296,144 cwts. were the produce of the liners (less by 95,118 cwts. than in 1892), and 425,687 cwts. the quantity landed by trawlers (45,582 cwts. more than in 1892). While the liners, therefore, showed a marked diminution in their total, the trawlers showed a considerable increase.

When the returns, however, of the fishing-boats of all kinds (other than beam-trawlers) are considered, it is found that there were in 1893, 363 fewer boats and vessels than in 1892, and a decrease of 1689 fishermen and boys. This condition of things is sufficient to account for a considerable diminution of line-caught fishes, without regarding the unfavourable weather of the season. Moreover, it has to be remembered that fishery statistics are far from being complete, for though the returns show that the quantity of fishes mentioned has certainly been landed, they do not indicate those fishes which have been landed and not reported. On the other hand, the number of the trawlers has increased by 2 (probably powerful steam-vessels) and 18 men during the year.

In 1894 the liners brought to land 1,250,066 cwts. of round fishes = £457,798, an amount exceeding the captures of the previous year by 113,677 cwts., yet the increased quantity was accompanied by diminished value, for the sum received was £8,601 less. The greatest increase was in perhaps the most important fish, viz. the haddock, which but lately had been the subject of many pessimistic theories. The captures exceeded those of the previous year by no less than 110,000 cwts., but, from the small size of the majority, the price had decreased by £3,389. The vital point, however, is the great abundance of small haddocks along the eastern shores in 1894. The variability both in amount and value was further shown by the fact that saithe (green cod) also exceeded the quantity in the previous year by 21,731 cwts., but was diminished in price by £1,374, while whiting was also in excess by 10,362 cwts., but the value was only £586 more. Cod, again, had decreased to the extent of 3,225 cwts., yet had an increased price of £1,751, ling was also less by 20,773 cwts., and fell behind no less than

£7,807, and conger showed a smaller diminution also in both quantity and value.

The trawlers produced 347,827 cwts. of round fishes = £168,136, an increase of 37,965 cwts., and a diminution of £10,168 on last year's captures If no other returns than those of the liners had been available the irregularity of the captures of the various forms might have been misunderstood, *e.g.* that the cod was greatly diminishing. The trawl, however, procured 24,840 cwts. of cod more than the previous year, and an increase of £4,054; 17,631 cwts. of haddocks more than in 1893, but with the great diminution of £11,883, showing, as in the former case, the prevalence of the smaller forms. Other kinds of round fishes showed considerable diminution.

The liners landed 105,244 cwts. of flat fishes = £59,734, nearly double the amount (48,095 cwts.) of the previous year, though the value was only £16,428 more, probably from increased abundance and diminished prices. This total is divided between ordinary flat fishes (pleuronectids) and skate. The former amounts to 61,783 cwts. = £49,133; an increase of 4,634 cwts. and £5,827 on 1893. Halibut, flounders, plaice and brill account for the increase, the larger share of which (£3,143) falls to the former, while £2,783 cover the three latter. Of skate the total for the year is 43,461 cwts. = £10,601, a diminution on the previous year of 9,165 cwts and £124. Of other kinds of fishes 32,149 cwts. = £7,808, or 14,311 cwts. and £2,918 less than in 1893.

The trawlers caught a less amount of flat fishes than the liners, viz. 77,660 cwts., but the value was much greater, viz. £91,887, a somewhat larger amount by 1,253 cwts. and £1,091 than in 1893. The total for flat fishes proper was 71,882 cwts. = £90,760, an increase of 857 cwts. and £979 on the previous year. The forms in which an increase occurred were flounders, plaice, and brill 1,099 cwts. and £2,208, and lemon-dabs 667 cwts. and £1,871. A diminution of 996 cwts. and £3,327 in turbot was the most important on the other side. Skate showed a slight increase, viz. 395 cwts. and £112; while other kinds of fishes, as in the case of the liners, were reduced by 16,536 cwts. and £3,621.

The value of the round fishes in 1894 is 7*s.* 3*d.* per cwt. for the liner and 9*s.* 8*d.* for the trawler, a balance of 2*s.* 5*d.* in favour of the latter. This is somewhat less (by 8½*d.*) than in 1893. For flat fishes on an average the liner received 11*s.* 4*d.* (1*s.* 6*d.* more than in 1893), the trawler 23*s.* 7*d.* per cwt. (2*d.* less than in 1893). Both had the same price for the other kinds of fishes, viz. 4*s.* 10*d.* per cwt. The disproportion in regard to the price of the flat fishes has always been a marked feature in contrasting the two classes of fishermen, and is probably due to turbot and lemon-dabs.

The grand total of all kinds of fishes landed in 1894 was 6,188,774 cwts. or less by 19,244 cwts. than in 1893. If herrings, sprats, sparlings and mackerel (viz. 4,351,628) are deducted 1,837,146 cwts. are left, an increase of 115,315 cwts. on 1893. Of this about 1,388,777 cwts. pertain to the liners or 92,633 cwts. more than last year, and 448,369 to the trawlers, or 22,682 cwts. of an increase over 1893.

This year (1894) also a decrease occurred in the number of boats propelled by sails or oars of 213, of fishermen and boys 721, so that this must be taken into account in considering the quantities landed by the liners. There was an increase, however, in steam-liners of 6, the numbers being for 1893—38, for 1894—44; and an increase of 50 men. The energy necessary for working steam-fishing vessels with profit has also some connection with the increase in this department.

During the year 5 steam-trawling vessels were added to the fleet (4 to Aberdeen and 1 to Leith) and 52 men. The total number in Scotland being thus 77, that is, trebled since the agitation on the subject commenced in 1883. Moreover the quality of the vessels has considerably improved. Besides these, 38 steam trawlers, other than Scottish, fished off the east coast of Scotland.

The liners captured in 1895, 1,343,011 cwts. of round fishes = £482,820, an increase of 107,065 cwts. and £25,022 on the amount and value of 1884. This considerable increase was largely due to the fact that 110,537 cwts. of haddocks more than in 1894 were secured, a point of some importance in view of the persistent efforts in certain quarters to pronounce the doom of the haddock all along the eastern shores of Scotland. Yet the greater part of this increase in haddocks, or 92,954 cwts., came from the east coast, and by no transference of the increment to the Great Fisher Bank, the Faröe Islands, or similar distant areas, can this be under-estimated. It is sufficient to point to the captures of the "Garland" within the limits to show that the haddock, if persistently followed, was likewise plentiful inshore. The substantial increase was not only in numbers, but the value rose by £34,756 on the total catch on the Scotch shores in 1895. Such facts strengthen the belief that the Scotch haddock fishery is not so reduced as is stated. It is true that every season, indeed every week, shows its variations —from weather, roving of the shoals of fishes and other causes, *e.g.* the use of unsavoury bait, yet after trawler and liner have each taken as many as possible the remaining stock is such as to give no cause for alarm—even to those sensitively alive to the welfare of the marine fisheries. The large number of boats and the use of trawls possibly disturb and break up the shoals for a time, as well as cause them to change their ground, but the ocean is a vast place of refuge—in which the depleted ranks will be augmented, their places taken by others, or by themselves at a later stage. It cannot be supposed that all the manifold apparatus persistently used by liners and by trawlers makes no impression on the hordes of haddocks. The author has always thought that the larger forms were diminished in numbers, and this is borne out by the appearance of the captures, say off Shetland on the one hand, and off St Andrews Bay and the Forth on the other—except under special circumstances as, for instance, the presence of shoals of herrings at their spawning season in winter. But to aver that, because the harassed and more wary fishes are caught in small numbers, they are on the road to extinction, is to shut out the facts that the water teems with the eggs in season, with swarms of minute young food-fishes, and with myriads of other young fishes which along with pelagic marine organisms of a lower type form their food. Eggs of haddocks in the water are as good hostages for the filling up of the ranks as need be wished; while the myriads of young in the deeper water, free from molestation by any kind of commercial apparatus, still further act as a bulwark against extinction. Everything is on the side of nature in the struggle. Even the capture of the adult spawning haddocks is not the final stage, for the decks of the vessels are often covered with masses of ripe eggs and milt which even the most careless manipulations in throwing or washing overboard cannot altogether sacrifice. More romantic than the Gorgon's head are the never failing resources of nature in connection with these and other food-fishes.

Considerable increase had also occurred in the captures of cod, which showed 4,969 cwts. and £1,452 over that of last year. There was a smaller increase in the quantity of torsk, conger and whiting, but the value of the latter was £1,112 less than in 1894. On the other hand ling had receded by 6,364 and £8,042, green cod (saithe) by no less than 19,546 and £2,853.

The trawlers also produced a larger amount of round fishes than in 1894, viz. 434,129 cwts. = £174,380, the increase being 86,302 cwts., but only £6,244 more than last year, which had a deficit of £10,168 on the previous year (1893). As in the case of the liners the captures of haddocks make a large proportion of the increase, viz. 78,389 cwts. and £2,675, showing that the size of the haddocks had been small, or that the values had been low. Cod show an increase of 6,603 cwts. and £4,032, and there is also a slight increase in ling, whiting, and a slight diminution in saithe.

Of flat fishes the liners landed 111,011 cwts. = £59,767, or a slight increase of 5,767 cwts., the value remaining almost the same as last year. Their chief increase in weight was in halibut, viz. 2,709 cwts. over the previous year, but the value had been diminished, since the trifling addition of £318 only resulted. Flounders, plaice and brill were less by 2,305 cwts. and £1,675. A large increase in skate was noteworthy of the year, viz. 10,877 cwts. = £1,778, yet this fish is as liable to be captured in the trawl as any other form, though, as it happened, the captures by trawlers were under those of 1894, one of the irregularities so prevalent in marine fishing. Of other kinds of fishes there was a diminution on the returns of the previous year, viz. of 157 cwts. and £6,517, showing that the market-values had likewise altered.

The trawlers captured 79,835 cwts. of flat fishes = £112,749, a less quantity than that caught by the liners in weight by 31,176 cwts., but exceeding it by the large sum of £52,982, that is, their flat fishes were nearly twice as valuable. The total quantity exceeded by 2,175 cwts. that of 1894, not a large increase, but fully maintaining the position, especially as it was accompanied by an increment of £20,862 in value. The total of flat fishes proper was 74,927 cwts. = £111,652, exceeding that of 1894 by 3,045 cwts. and £20,892. The cost of flounders, plaice and brill had increased, since with a surplus of only 860 cwts. over 1894 the value was raised by £11,202. Lemon-dabs had an increase of 1,440 cwts. and £7,153, and 556 cwts. and £2,105 were added to the totals of the turbot of the previous year. A slight increase occurred in halibut, whereas the liners had fallen 2,709 cwts. behind the returns of 1894. Skate, again, were less by 870 cwts. and £30; while other fishes, as in the case of the liners, had diminished by 5,151 cwts. and £5,151.

The value of the round fishes in 1895 is 7*s.* 2*d.* per cwt. for the liner and 8*s.* for the trawler, the difference (10*d.*) between the two kinds of fishermen being less than the previous year (2*s.* 5*d.*). In the case of flat fishes the liner received 10*s.* 9*d.* per cwt., the trawler 28*s.* 3*d.*, the disproportion (17*s.* 6*d.*) being greater than in 1894, and due to high prices as well as to increase in turbot and lemon-dabs.

The grand total of all kinds of fishes (other than shell fishes) landed in 1895 was 6,107,044, a total less by 81,728 cwts. than in 1894, but this decrease does not affect the liner and trawler, since if we deduct herrings, sparlings and mackerel (viz. 4,095,695 cwts.) 2,011,349 cwts. caught by liner and trawler are left, a quantity exceeding that of 1894 by no less than 174,203 cwts. Of the total thus captured, 1,479,654 cwts. pertained to the liner, and 531,695 cwts. to the trawler, the former showing an increase on the previous year of 110,877 cwts., the latter of 83,326 cwts. Yet these results have been attained by a diminished number of fishing (sailing) boats, viz. 194, and

523 men, their places having been taken by two steam liners and 25 men. The statistics show that the supply of fishes for the market has notably increased in this department, notwithstanding the steady decline in the numbers of those working sailing boats, the men whom we for generations have associated with this perilous industry, and who are noted for their hardihood and daring. This change has been largely brought about by the introduction of steam, both in line-fishing and in trawling.

No steam-trawling vessels were added during the year, a feature of interest in connection with substantial increase in the captures. The tonnage, however, had an increment of 53 and the value had risen by £13,430. Only 5 additional men were engaged in this department. There were, however, 3 deducted from the list of other than Scottish vessels (38) fishing off the east coast of Scotland.

In 1896, the liners procured 1,449,259 cwts. of round fishes = £508,928, an increase of 106,248 cwts., and no less than £26,108 on the previous year. Thus year by year the steady increase in this important industry continues. Last year the haddock had the credit of raising the total in a noteworthy manner. This year the cod is in that position, by an increase of 90,416 cwts. and £19,751 over 1895, and the green cod (saithe) by about half the amount in weight. The quantity of haddocks still more than maintained its position, for there was an increase of 7,137 cwts. and £10,946, a satisfactory result for two years in succession. The increase in the green cod (saithe) was 46,398 cwts., but only £2,953, a fact due to the moderate price always attached to this fish. Conger had increased by 4,463 cwts., but the value was £221 below that of 1895. On the other hand, ling had receded by 37,973 cwts. and £7,742 and whiting by 1,081, with an increase in price of £446. That for two years the ling fishery had so markedly declined was probably due to a change in the methods of the fisherman, or an absence of the previous energy.

The trawlers likewise showed an increased quantity and value of round fishes over 1895, viz. 444,250 cwts. and £178,604. The actual increase was 10,121 cwts. and £4,224. While cod, as in the case of the liners, showed a decided augmentation of

25,322 cwts. and £4,251, and saithe (coal fish) of 2,953 cwts. and £228, the other round fishes varied. Thus whereas in the liners a great deficit occurred in ling, the trawlers had 2,758 cwts. and £507 more than in 1895. Haddocks, however, had fallen behind 1895 by 18,279 cwts. and £67. Whitings were barely on a par with the previous year, but congers maintained their position. Such variations are inseparable from sea-fishing, and show how necessary it is to be cautious in drawing deductions either from statistics or the statements of those interested. Some of these variations have great antiquity, and are even indicated by the seasonal pursuits of the fisherman, who when the haddock fishing becomes less tempting in June takes to the herring fishing in distant waters, and the successful financial result as well as the healthy variety of occupation cannot but have an important influence on his happiness and prosperity. Plaice-trawling or fishing in late autumn, on return from the herring fishing, is another example. Others are the result of chance, and are exemplified in the casual capture of cod in a stake-net, while the liners see little of them, by the loading of one boat's lines while another has few haddocks, by a sea swarming with whitings, haddocks or herrings, while for a long period all three may be comparatively scarce.

In flat fishes the liners also held a better position than in 1895, the total being 113,103 cwts. and £67,034, or an excess of 2,093 cwts. and £7,267, the latter disproportionate increase in value being due to higher prices. This increase was made up of 143 cwts. and £223 for turbot, and 4,171 cwts. and £5,885 for halibut. Lemon-dabs remained nearly stationary, while there was a loss by weight of 2,813 cwts. and an increase of £52 in flounders, plaice and brill. Skate maintained the position of 1895 by having 598 cwts. and £1,114 over that year. Other kinds of white fishes had an increase of 1,210 cwts., but a diminution of £258 in value. The great increase in value of last year's flat fishes must be borne in mind when considering the further increment this year. Altogether the results are satisfactory.

The trawlers obtained 90,652 cwts. of flat fishes = £124,495, an increase of 10,817 cwts. and £11,746, thus maintaining steady progress. The important turbot showed 1,683 cwts. and £2,022

Wife of fisherman carrying home line in basket ("scow") on the arrival of the boat.

To face p. 90]

over 1895; halibut remained nearly as in the previous year, while lemon-dabs had a diminution of 733 cwts., but an increase of £160 in value. Flounders, plaice and brill exceeded 1895 by no less than 7,973 cwts. and £8,502. Skate had increased by 1,888 cwts. and £340. Other kinds of white fishes also had an excess of 2,110 cwts. and £1,245 over the previous year.

The value of the round fishes in 1896 is 7*s.* for the liner and 8*s.* for the trawler, the difference in favour of the trawler when contrasted with 1895 being 2*d.* The liner received 11*s.* 10*d.* for flat fishes, the trawler 27*s.* 5*d.*, an increase of 1*s.* 1*d.* in the former case and a diminution of 10*d.* in the latter.

The grand total of all kinds of fishes (other than shell-fishes) landed in 1896 was 6,146,738 cwts., an increase of 39,697 cwts. on 1895, and this notwithstanding the great deficit in the herrings, sparlings and mackerel of 92,904 cwts. on the previous year. If we deduct the total for herrings, sparlings and mackerel, viz. 4,002,791 cwts., the fishes caught by the liner and trawler amount to 2,143,947 cwts., an excess of 132,598 cwts. on the previous year. Of this total 1,589,204 cwts. were produced by the liner, and 554,743 cwts. by the trawler, the former presenting an increase of 109,553 cwts. and £33,117, the latter 23,048 cwts. and £39,697. This year also a large reduction of boats—boats propelled by sails or oars—took place, viz. 1,064, representing 4,303 in tonnage, and 3,521 resident and non-resident men and boys. The remaining boats, however, were increased in value both as regards the hull and equipment. Besides, 7 new steam-liners and 93 men were added to this department, a sign that the old order of things is steadily being dispensed with.

No steam-trawlers were added to the Scottish series this year, and a reduction of 3 in vessels other than Scottish took place. Two small sailing trawlers were also added to this department. The increase in men was 31, and the tonnage in the steam vessels was augmented. The general adaptation of the otter-trawl in the trawling vessels was a feature of the year, indeed, so great is the competition in the fishing industry, that it was found that the beam-trawls could scarcely hold their own in the race. Hence the new system invented by Mr Scott, of the Granton Steam Fishing Company, became a necessity.

The year 1897 was marked by a considerable diminution in line-caught round fishes, the total being 1,276,702 cwts. = £492,103, a reduction of 172,557 cwts. and £16,825 on the previous year. The amount was also exceeded by that of 1895, though it was higher than in 1894. The greater part was caused by a decrease of 154,758 cwts. of haddocks, valued at £26,053. This reduction chiefly affected the nearer waters, for the captures in the more distant waters by the trawlers were satisfactory. These nomad fishes had apparently been somewhat peculiar in their distribution. Cod had diminished by 3,990 cwts. but increased in value £2,615. Torsk likewise fell 2,158 cwts., and saithe no less than 34,584 cwts. Whitings were less by 9,274 cwts. and £1,696, but congers remained in the former position. On the other hand ling had substantially increased by 32,154 cwts. and £10,583. The line-fishing, therefore, appeared to have been less successful generally than usual.

The total for round fishes caught by the trawl was 499,259 = £229,506, an increase on the previous year of 55,009 cwts. and £50,902. It is interesting that this substantial increase was made up of 30,874 cwts. and £40,727 for haddocks over the previous year, showing that at least they were abundant somewhere. A decided increase also occurred in cod, viz. 19,679 cwts. and £7,759, and in saithe (green cod) of 2,484 cwts., whitings 3,606 cwts. and £2,357, while congers remained about the same. On the other hand there was a decrease of 1,641 cwts. in ling. The trawlers therefore must have been of signal service in making up the deficit of the liners in haddocks, and in demonstrating that this species holds its own in the sea. Moreover, the powers of reproduction in such a species as the haddock also make it certain that wherever a thinning has occurred the swarms of eggs and young from the neighbouring regions will by-and-by restore the balance at the usual seasonal period.

The liners had a total of 112,957 cwts. = £72,068 for flat fishes, a result only 146 cwts. under last year in weight, and exceeding the value in 1896 by £5,034. Little change occurred in regard to turbot, but halibut had a decrease of 3,450 cwts., though the value was £1,639 above that of 1896. Lemon-dabs showed a trifling decrease, as also did flounders, plaice and brill. Skate showed an increase of 3,379 cwts. and £3,188, and other kinds of fishes showed the slight diminution of 264 cwts., but an increase in money of £76. The year, therefore, was an average one, and the variations were probably in a large measure due to the weather.

Of flat fishes the trawlers secured 71,808 cwts. = £108,454, a considerable diminution on the totals of the preceding year, viz. of 18,844 cwts. and £16,041, a result, as regards weight, less than any year since 1889, the period at which the statistics began. The value, however, has been considerably enhanced, and therefore the total is only less than the two preceding years. Turbot were less than in the preceding year by 410 cwts. and £540, and lemon-dabs by 5,355 cwts. and £5,989, while flounders, plaice and brill showed the serious diminution of 14,156 cwts. and £10,299. There was a large increase last year on the results of 1895 under this head, so that the reduction was comparatively sudden. Whether, as suggested in the Report of the Fishery Board, the new otter-trawls capture more round than flat fishes, or there has been a lack of attention to this department remains to be seen. Certainly in inshore waters plaice-fishing has not been prosecuted during the year with the usual energy, but whether this was the result of scarcity of fishes or pre-occupation with more remunerative work is unknown. The closure of the inshore sandy areas places the supply of plaice from 7 to 13 inches or more in the hands of the liners, and there can be no question as to the great diminution of this class of food since the waters were closed. The liner can capture a considerable number, but only by the exercise of energy, and he has the market in his hands. Lately a slackness of operations in the inshore (closed) waters has been apparent, though these have been specially reserved for the benefit of the liner. The boats everywhere push out beyond the limits to compete with the trawler in the open sea; while the smaller inshore fishes are left to themselves. So far as the operations of the "Garland" can be relied on, and they are trustworthy in regard to St Andrews Bay, there are as many flat fishes in these inshore waters as before the experiments began, and there does not appear to be any cogent reason why the public should not benefit by their presence. The sending of a powerful and efficient trawler into such bays now and then would have been a valuable test of their condition, and would have enabled the Fishery Board to arrive at conclusions of considerable practical importance.

The value of the round fishes in 1897 was for the liner 7*s*. 8*d*. and for the trawler 9*s*. 2*d*., the difference in favour of the trawler when contrasted with the condition in 1896 being 6*d*. The liner received 12*s*. 9*d*. for flat fishes, the trawler 30*s*. 2*d*., here again the increase mainly falling to the trawler, whose total (30*s*. 2*d*.) was 2*s*. 9*d*. over that of the preceding year.

The grand total of all kinds of fishes (other than shell-fishes) landed in 1897 was 5,001,672 cwts., a diminution of no less than 1,145,066 cwts. on the previous year, yet the aggregate value was £55,951 over the captures of 1896. This reduction was due to the small catch of herrings, which fell by 1,010,701 cwts. under that of the 1896, yet the quality of the fishes and the prices obtained actually gave their captors a surplus of £32,390 on this head alone over 1896. The total caught by the liners and trawlers (after deducting the herrings, &c.) is 2,009,582 cwts., a decrease of 134,365 cwts. on 1896, that is to say, this deficit removes the excess of 1896 over 1895. The share of the liners in this total fell short by 172,967 cwts. and £11,715 of the captures of the previous year (1896) but the trawlers kept advancing, as they have done since 1884, their increase this year being 38,601 cwts. and £35,276. There is reason to believe that but for the activity and energy of the steam-liners, the line-fishing would not have held in later years so good a position in comparison with the trawlers, whose total was 593,344 cwts. = £343,656. A further reduction of 430 boats propelled by sails or oars took place, and 1,049 resident and non-resident men and boys—connected with the department. An improvement, however, of the value of £10,767 took place in those remaining. On the other hand 4 steam-vessels and 37 men and boys were added to the liners, which thus are steadily increasing. Year by year, therefore, changes in the methods of line-fishing are progressing, such as steam-liners, and steam-power for hauling nets and lines.

Nine steam-trawlers and 6 sailing trawlers and 104 men and boys were added to the fleet, and considerable improvements in the construction and accommodation were made in the best vessels.

In 1884 trawling was carried on within a 'reasonable distance' of land, so that the paddle-ship could deliver the catches of the night in time for the market next morning, or the daily 'carrier' of the fleet of steam-trawlers from Granton, by leaving the fishing-grounds in the afternoon or evening with the united catch, could reach that port early next morning. The vessels from the Moray Firth could land their fresh fishes at Macduff or Aberdeen, and the vessels from Montrose and Dundee carried fresh fishes to those towns.

For fifteen years the trawl-fishery has been prosecuted with vigour, and it is interesting now to see what areas the ships frequent, and with what results. To commence with the most northerly, viz., Aberdeen, at which trawling has made great progress since the former date (1884), it is found that, whereas the chief supplies were brought fresh from the adjoining sea by the older paddle-ships, or from the Moray Firth by the more powerful vessels, much of the supply of the present day comes from the 'Great Fisher Bank' or from Iceland. Instead of the activity displayed in 1884 in the strip of sea from 10 to 20 miles off the coast, between Aberdeen and Montrose, only a few vessels are now seen at work here and there in good weather. Fishes are by no means absent from this area, and at certain times occur in considerable abundance, but the individual catches at other times are limited; and on the rough ground 10 or 11 miles off, in 33 fathoms, it sometimes happens that, after three hours' trawling, about the same time has to be spent in mending the net. Yet lemon-dabs and sail-flukes or 'megrims' (*Arnoglossus megastoma*) in the deeper and softer parts, with the larger haddocks and other forms, render the work there still worthy of attention. If small haddocks brought fair prices, the work would, indeed, be tolerably remunerative, as they are sometimes present in great numbers. The liners work on the same ground and catch chiefly the latter fishes. There is no indication that fish-food has been seriously interfered with on this ground, but, on the contrary, invertebrate life of all kinds is in great abundance. Moreover, the enormous numbers of pelagic sand-eels, from 15 to 33 mm. in length, intermingled with swarms of young flat-fishes, on these grounds, and on which many of the fishes were feeding in May, is a feature of moment. In 1884 the captures on the northern part of this area during the summer months were comparatively limited, and it was only the advent of the herring in autumn that caused a notable increase of white fishes. At the distance from land just mentioned, in 1894, each haul in daylight produced from a basket to a basket and a half of lemon-dabs, about three-fourths of a basket of large haddocks, and four to five boxes of small haddocks. At night, a few ling, cat-fishes and cod were added to the catch. Few whitings were procured, and the same feature was occasionally seen in 1884, for the whitings are often in the upper parts of the water. Very few cuttle-fishes occurred in May. The 'catch' just mentioned is not a heavy one, and is probably surpassed by other ships, but it at any rate shows that fishes are still present in considerable numbers. This is further demonstrated by examining the 'catch' of a liner with six men on board, and which had been at sea about 32 hours, fishing on the 28th and 29th of May, probably from 28 to 30 miles off Aberdeen, viz., 9 boxes of large haddocks, the largest fish reaching the length of 20 inches, the rest smaller (at 24*s.* per box), 3½ boxes of small haddocks, a few cod, dabs, one lemon-dab, and a few whitings—making a total of about £12 for the six men. In the same market lately the large haddocks brought 29*s.* per box, so that the above is probably not an unusual price.

At the southern end of the ground just mentioned, viz., off Montrose, a trawler working, in 1891, about 25 miles off, in August, landed the very high catch of 500 boxes of haddocks in a single night. In 1894 the takes ranged, per week, from 100 to 140 boxes of haddocks and flat-fishes, besides cod, coal-fishes, and gurnards. Plaice were said to be rather scarce, even lemon-dabs being more abundant. For the night of the 29th May 18 boxes of haddocks and flat fishes were landed, besides cod and ling. For each box of good haddocks (7 stones) 16*s.* were obtained, a much lower price than in Aberdeen, where, however, the box was heavier (8 stones or more). The 'catch' for the night was about a ton in all. A small liner, with five men on board, which went out between 9 and 10 A.M. on the 29th, landed at 5 P.M. (*i.e.*, in 8 hours) ¾ box of large, ½ box of medium, 2 boxes of small haddocks, many about 9 inches long; 1 lemon-dab, 2 very fine cod, and 4 codling, and this though their lines were 'shot' in broad daylight. The fishing-ground was from 8 to 10 miles off. This is a small 'catch,' but the circumstances under which it was made were not favourable. There can be no doubt that the entire eastern coast abounded with multitudes of small haddocks, and that these had been captured in immense numbers by both liners and trawlers.

The best trawling ships at Aberdeen, which were about 30 in number in 1894, chiefly frequent the Great Fisher Bank, about 200 miles off, and from 30 to 40 fathoms in depth, it being a general opinion amongst fishermen that this and up to 60 fathoms is the most favourable depth for their pursuit, for they think that in deep water (100—175 fathoms) they get only conger, halibut, and skate; and elsewhere, as off the coast of Portugal, only sharks are procured at 500 fathoms. Yet the Rev. W. S. Green, off the west coast of Ireland, got 'witches,' ling, haddocks, and conger at 170 fathoms, and skate and forkbeards at 500 fathoms. On this ground (Great Fisher Bank), which is about 120 miles from east to west, and from 60 to 80 miles from north to south (a larger area than the inclosed region of the Moray Frith), the 'catches' of these trawlers vary from 80 to 180 boxes or more, consisting of plaice, haddocks, turbot, and other fishes, which are procured in from 8 to 13 days, including the time spent on the voyage. Since the Moray Frith was closed, these ships, therefore, find it remunerative to undertake this long journey, and bring their fishes preserved in ice to the market in Aberdeen. They do not seem, however, to find it so profitable to fish in the waters near the Scottish shore[1].

[1] Some worked lately with success within 30 miles of Aberdeen. January, 1899.

In the same way, the powerful ships which proceed to Iceland bring from 200 to 400 boxes of fishes in about 14 days. The plaice procured in this region are recognised by the dark spots; and as these, the haddocks, cod, and other forms, have been 'gutted' and preserved in ice, they do not have so attractive an appearance as those caught by the liners.

Besides the areas just mentioned, some trawlers proceed to Blacksod Bay, off county Mayo, on the west coast of Ireland, for soles and turbot, while in February and March others go to ground 20 to 40 miles off Scarborough, where, perhaps, 20 score of cod are caught in a night. Some, again, work on the turbot-ground, from 80 to 90 miles off Aberdeen, and others find on the Dogger Bank catches of from 18 to 20 boxes of plaice.

When the trawlers from Granton and Leith are considered, it is found that, notwithstanding the closure of the Forth (for 3 miles beyond the Island of May), these ships have increased in number, have been improved in equipment, and have been able to overcome the difficulties with which they were handicapped —in comparison with the liners. In the case of the General Steam Fishing Company's ships, and probably in others, however, very definite instructions—based on carefully recorded data, compiled during the last twelve years—are issued to each captain as to the distance to be traversed (by the log), and the direction on every occasion. No haphazard selection of fishing-grounds is made. Thus in December, 1893, besides the ordinary fishes, numerous cuttle-fishes (so valuable for bait) were procured off the Isle of May. In January, February, March, and April they work from 5 to 10 miles S.E. of the Isle of May, viz., more or less on the grounds frequented in 1884. In March and April the cod are captured as before in considerable numbers as they congregate during the spawning season, and in the earlier months as they follow the herrings. In June, July, August, September, October, and November they take to the more distant grounds off the Forth—about 40 miles E. by N.E.

The opinions somewhat freely expressed by some in 1884 as to the decline of the trawling industry in the Forth and the adjoining area—notwithstanding all the advantages of a free area from inshore to offshore then possessed—do not seem to have been borne out by further experience. Even with the entire area of the Forth and St Andrews Bay closed, these vessels, now considerably increased in numbers, have found fishing profitable on the more distant grounds. They work on a certain area, either by means of a flag-buoy or otherwise, and strictly in accordance with the instructions given from head-quarters. If the captures are observed to be decreasing, either from the thinning of the fishes or their being scattered, they change ground, as, indeed, was very noticeable in 1884, returning after an interval to the same area, to find that an increase has taken place. In connection with this filling up of areas over which trawling has been assiduously carried on, it is an interesting fact that the local boats—from 12 to 20 or more in number—found for many years that, on the whole, their best ground in St Andrews Bay was a line about 2 miles from shore ('Scooniehill,' in a line with 'the steeples'), and about 4 fathoms in depth. Boat after boat trawled along that line, wind and weather permitting, for four months of the year, and sometimes longer, and to the closing day it maintained its position as the best area for plaice. The same observation has been made at Brixham, where trawling has been in operation about a hundred years. It is quite evident, therefore, that other fishes took the place of those captured, and that this continued month after month, and year after year. The whole question, consequently, in the larger areas outside the 3-mile limit is—Can the supplies from the neighbouring waters keep pace with the rate of capture now going on by both liners and trawlers? These supplies consist of the growth of the young from eggs on the area itself, and the immigration of eggs, young, and adults from other areas, or the open water beyond. It is seen that, so far as human observation can go, the supplies of herrings are as plentiful as formerly, notwithstanding the absence of restriction and the great waste that annually takes place in this fishing. On the other hand, it is a matter of observation that the first hauls of the trawl on virgin ground are the most successful, and that by and by the catch diminishes, and the same occurs with the liners on their new 'banks' or 'reefs.' Yet it cannot be said in either case that the fishes have been extirpated, but they probably have become more wary as well as diminished in numbers, and, moreover, they may have changed their ground, for fishes are constantly roaming. It has to be remembered that the food-fishes are not altogether confined to the shallower water, in which they are usually followed, but they likewise extend into the deeper water beyond. Such deeper water and unfrequented regions, therefore, form reserves, in which the species is reproduced, the eggs, young, or adults passing into those areas in which the food-fishes have been more or less thinned.

The area last mentioned, viz., that off the Forth, is perhaps one of the most important in Scotland in regard to the number and variety of its fishing-grounds. For the present purpose the area may be defined as that bounded by a line drawn eastward from Arbroath on the north, and a similar line from St Abb's Head on the south. Between these points the Tay and the Forth pour considerable bodies of fresh water into the sea, while the Eden debouches into St Andrews Bay between them. The amount of microscopic food—both plant and animal—as well as of the smaller invertebrates which are carried to sea in this area, is very considerable, and, in all probability, is closely related with the richness of invertebrate life both in the waters and on the bottom. The enormous numbers of pelagic mussels swept from the Tay and the Eden alone form a remarkable feature. It is not surprising, therefore, that the fishing-grounds in this region still continue fairly prolific, notwithstanding the increased demands on their resources. In the same way the Moray Firth is another rich fishing-area on the east coast, though the rivers entering it are smaller.

The steam-liners and trawlers frequent the more distant grounds—not because the fishes are absent from the nearer grounds, but because their 'catches,' as a rule, far exceed in bulk those obtained on the latter. While, therefore, the present statistics show no serious diminution, it may be truly said that the total is kept up only by the supplies from Iceland, Faröe, and the Great Fisher Bank. But the nearer grounds would have produced a considerable supply if they had been perseveringly worked; and it cannot be doubted that they contained, at any rate in 1894, an immense number of small haddocks[1]. Moreover, these small haddocks had migrated from the distant waters, for it is a remarkable fact that, so far as ascertained, no great shoals of very small haddocks (*i.e.*, less than 3 inches) have been encountered in inshore waters. The life-history of the haddock, indeed, shows that between its post-larval condition and the adolescent stage of between 2 and 3 inches, it is a deep-water fish. Before the appearance of these hordes of small haddocks, it was generally asserted that the haddock had been more or less extirpated; hence the necessity for caution in dealing with such subjects. Again, the question as to the completeness of the statistics of fishes caught round the Scottish shores has to be considered, and there are some who think much improvement is required in this direction. Indeed, the only satisfactory method would be for every liner, trawler, net-, crab-, or other fisherman to hand to the official on reaching the port a slip stating the amount and kind of the 'catch,' and the ground on which it was made, as indicated in the Trawling Report of 1884.

The condition of the inshore waters (within the 3-mile limit) has elsewhere been dealt with[2], and will again form the subject of future remarks. All that need be said at present is that, so far as can be ascertained, it would not appear that the closure of the inshore waters has made any marked increase in the fishes of the offshore waters, yet the younger fishes have now had time to pass outward and become mature; nor have the larger fishes been driven shorewards by the more frequent interference with the more distant areas. No change, however, could be expected if the scarcity were due to general over-fishing.

A consideration of all the foregoing facts gives no grounds for despairing of the sea-fisheries—but sifted in every way show that the condition is not unsatisfactory in regard to the

[1] An idea of the numbers of these may be given by stating that a trawler brought on board, in two hauls, about ten tons of small haddocks, which were, however, freed. Many were probably killed.

[2] *A Brief Sketch of the Scottish Fisheries*, 1882—1892, p. 6.

supply. The only serious question that appears is the steady diminution in the number of the sailing-boats and of the men fishing in them. Several factors probably are connected with this change—the first is the abundant supply brought in by the steam-trawlers and steam-liners to great centres in easy communication with railways and a large population, so that the prices paid to outlying towns and villages less favourably situated are greatly reduced and even rendered uncertain.

The number of powerful ships with great trawls probably tend to scatter the fishes, and render them more wary in regard to all kinds of capture, and it may be to keep further from land, and while the fishes may not, taking a broad survey of them, be very much reduced, and the totals in the market even increase, yet each ship and boat probably secures much less than in former times.

In any case, the thoughtful perusal of the foregoing annual statistics of fishes caught by line and trawl does not give rise to dissatisfaction, and, taken with the results of the work in 1884, and the ten years' experiments of the "Garland," together with other observations, conduces to confidence in the future of the sea-fisheries.

Aberdeen Fish-Market at 5 a.m., August, 1898. The boats land on the right, and carts remove on the left.

To face p. 102]

CHAPTER III.

Scientific Investigations in St Andrews Bay. 1886—1896.

Tables I.—VII.

The Stations in St Andrews Bay were as follows:—

Station I. extends 5 miles S.E. by E. from St Andrews, depth 6—12 fathoms.

Station II. begins 2 miles from St Andrews, and extends E. by S. for 4 miles. Depth 7—14 fathoms.

Station III. begins 2 miles from the pier, and passes 4 miles E. by N. 8—12 fathoms.

Station IV. commences a mile and a half from harbour and curves round the N.W. part of the Bay. Depth not exceeding 8 fathoms.

Station V. is outside the enclosed area, beginning 2 miles from the Tay-buoy, and ends E. half S. of Barbet Ness[1].

Station VI. runs nearly at right angles to the latter, or from the Carr water towards the Bell Rock.

In examining the statistics carefully sent from the "Garland" by Mr W. L. Calderwood[2], the naturalist on board, and compiled by Mr Duncan Matthews at the Fishery Board, the mode of drawing deductions from them has been varied; in short it has been deemed well to handle them in the same way as was done for the Report of the Trawling Commission in 1884, viz., by

[1] *5th Ann. Report, S. F. B.*, p. 55, 1887.

[2] Now H. M. Inspector of Salmon Fisheries for Scotland.

dividing the fishes captured into two groups, saleable and unsaleable (chiefly immature). It has to be borne in mind that during the first season or two, from insufficiency of funds, the "Garland" was not kept at work the whole year. In 1866, the work was carried out only during the months of June, July, September, and November, and it is possible that the trawl was less effective than it subsequently became. In the Tables, moreover, barren hauls are omitted[1]. During the months mentioned seventeen hauls of the trawl were made in St Andrews Bay, an area in which there were five stations, one (V.) beyond the limit, and the rest within it, and about which a considerable amount of information has been collected during the last forty years. A total of 1,792 saleable fishes were procured, the majority being plaice, which were mostly captured within the limits, though in June and September a considerable number were got on the outer station (V.), where four out of the six larger were also caught. Gurnards follow, the largest number of these occurring at the latter station. Dabs are next in order, and from the four inner stations, then haddocks—dispersed generally over the area. Long-rough dabs were likewise generally distributed, though the largest number (21) were obtained in November at the station beyond the limit. Two large lemon-dabs were got at the latter station, and fifteen of the twenty smaller in the same line. Only two large cod, one large haddock, two large whitings, a single coal-fish, four flounders, a single turbot, a single brill, and a sprat were obtained in the area; eight skate were likewise procured.

Of unsaleable (chiefly immature) fishes there were 1,186, more than half being dabs, which were generally distributed over the area, the largest catch, however, having been made in September on Station III. in the middle of the bay, and the next in order on Station I. along the southern border—during the same month. The plaice (277) were likewise generally distributed, the largest number having been got in September on Station IV. along the inshore sandy margins of St Andrews and Abertay. The small gurnards were all pro-

[1] This refers to the tables alluded to in 1892, and to the annual tables from which those in this work were made up.

cured within the limits of the bay, and so were the long-rough dabs, but a few flounders appeared on the station (V.) beyond the limits. The small cod were all got in November, and were generally distributed; and the small haddocks were procured in September and November, probably on their passage shorewards. A single turbot and one whiting were obtained. A few skate (17) and one or two frog-fishes and skulpins were in the list.

Calculated on the total hauls (17) the average of the saleable was 105 fishes, and the unsaleable 69 or a total of 174.

This therefore formed the basis for future observations in St Andrews Bay, but it might well have been more exhaustive, since the conditions immediately after the cessation of steam and other trawling had to be determined. Fortunately, however, records made in 1884, either for purposes of trade or for scientific experiment, are available. Thus four hauls of the trawl (which was over fifty feet in length) were made by the powerful iron ships of the General Steam Fishing Company, Granton; two by fishing-boats of the place with the ordinary trawl of 30-foot beam; and fourteen by the "Medusa," the little steam vessel[1], from the Granton Laboratory with a trawl-beam of ten feet, and a net of small mesh.

It has also to be remembered—in considering the work of the "Garland"—that the results obtained by small trawls, even with a net of the same mesh as the larger, deviate considerably from those of a beam of fifty-six feet raised about 4 feet from the bottom. Thus the large fishes, which often escape such a trawl as that of the "Garland," are captured in numbers by the large net of the trawlers from Granton. This will at once be apparent by contrasting the captures of the more active round fishes, such as the cod and haddock, in the respective ships; and further by the absence, as a rule, of such a fish as the herring from the returns of the "Garland." Further, the work of the "Garland" was for the most part carried on in daylight, whereas the commercial ships utilized continuously night as well as day, and their whole energies were concentrated in their pursuit. As certain fishes are captured more readily at night,

[1] Most courteously sent by Sir John Murray.

and the average catch by this net is improved in darkness, the commercial ships had an advantage. The returns of the "Garland," therefore, from the beginning have to be considered to some extent on their own merits. Instead of continuing the work of 1884, extending it, and filling in gaps, new experiences, of less moment in the question, were created, and the comparison of results rendered more complicated.

In 1884 the four hauls by the powerful trawlers from Granton took place in January and May, and as subsequent experience in the "Garland" showed, these two months were not characterised by very high averages. They yielded 2,521 saleable fishes, composed early in January of 402 haddocks, 6 large cod, 34 plaice, 24 whitings, and 18 dabs, besides other forms in small numbers such as thornbacks (which are absent from the returns of the "Garland"). About a fortnight later in the same month, the saleable round fishes were still more abundant, viz. 98 large haddocks, 639 small haddocks, 122 whitings, 77 plaice, besides cod, turbot, brill, grey skate, thornback, and long-rough dabs. The other two hauls were made on the 1st and 2nd of May in daylight, and it is interesting to observe the change that had occurred in the fish-fauna. Taking the two hauls together for brevity, it is found that the first place is held by the more or less ripe grey gurnards, which numbered 575, the next by the plaice, viz. 372; flounder 70, haddock 14, long-rough dab 9, thornback 6, grey skate 7, turbot 7, sole 1.

Of the unsaleable fishes there were at the beginning of January:—angler 6, whitings 8, thornback 3, dab 5; while a fortnight later the proportions were angler 3, long-rough dab 47, dab 12. In May the two hauls gave—angler 7, grey gurnards 61, long-rough dab 96, dab 136. While these fishes would have been unsaleable at Granton, they would in most cases have found a market at St Andrews.

It would seem that most of the round fishes of the cod-family had left the bay, while an immigration of grey gurnards had taken place, but it is well to remember that the very stormy weather in May may have sent the gurnards downward in the water, and thus brought them more readily within reach of the trawl. Whatever the cause may have been the facts are interesting.

The two hauls by a fishing-boat with a 30-foot beam were made in daylight in June, and produced 215 saleable fishes, chiefly plaice, dabs, flounders, and thornbacks (36), the large number of the latter probably being due to the close approach made to the shore (sand). The smaller and mostly unsaleable fishes (182) were chiefly plaice, dabs, and gurnards.

Fourteen hauls, chiefly in daylight, by the "Medusa" with a small meshed trawl of 10-ft. beam gave in July and August 514 saleable fishes, the majority being plaice and dabs. Only two haddocks and five gurnards represented the round fishes, which were too active to be captured by a small trawl. Considerable variety, however, was shown, for turbot, brill, thornbacks, long-rough dabs, lemon-dabs, and flounders were represented. The unsaleable food-fishes were 833—a comparatively limited number when the small meshes of the naturalist's trawl are considered, though quite out of proportion when contrasted with the same class in the commercial ships. They consisted mainly of dabs, plaice and gurnards, the largest number being captured in August, in which month a young haddock, a few poor cod, whitings and coal-fishes were included. There were fourteen examples of other fishes, such as frog-fishes, skulpins, pogges, &c.

From the foregoing it would appear that St Andrews Bay was, throughout, fairly supplied with food-fishes towards the close of the open period, and that the enormous numbers of small plaice and dabs formed one of the salient features. The nomad fishes, such as the gadoids, gurnards, and herrings, were more or less seasonal in their occurrence, the former especially characterising winter, a single boat sometimes capturing from 50 to 80 cod within the limits of the bay.

But in addition to this method of estimating the piscine contents of the bay, an examination of the rocky margins and of the tidal pools showed that in July swarms of young cod and young pleuronectids occurred, the former, gradually increasing in size, remained till late in autumn, while the pleuronectids distributed themselves along the sandy margins and into the estuaries.

In 1887 a Committee of the Scottish Fishery Board in referring to the progress of the scientific work gave a preliminary outline of the methods by which they hoped to obtain information and such was chiefly based on the advice they had received.

In treating of the results of the year's trawling by the "Garland" in the closed area of the Forth and just beyond it, the reporters observe[1], "Conversely, while there has been a general advance in the numbers of flat fish at all the stations in both series, the rise at the inner stations has been by far most marked, the average increase here in 1887 being 128 per shot, while at the three outer it was only 22 per shot." They further state "that at the outer series of stations the increase has been greatest in lemon soles, next in plaice, and least of all in dabs. At the inner group, on the other hand, the augmentation has been chiefly in plaice, closely followed by dabs, and at some distance by lemon soles." At the two stations in the free waters (stations 8 and 9) "plaice and lemon soles have increased more than dabs."

In St Andrews Bay the increase was found "to be even more marked than in the Forth." They state that, as might have been expected from the physical characters of St Andrews Bay, "the increase has consisted principally of flat fish, although the round fish have also increased to a very considerable degree. ...Cod, which is rare in the bay, diminished; whiting remained about the same, but there was a large increase in haddocks and also in gurnards." The station in the free waters outside showed a diminution in plaice, increase of lemon-dabs, dabs and skate. Whiting have diminished, and haddocks and gurnards have increased. The reporters mention that they have had several instances of the "replacement of the round fishes by the flat, and *vice versa*."

The trawling experiments in Aberdeen Bay, a station specially selected as one on which a large amount of trawling had been carried on, were stopped. It would have been important to have noted the changes in the fish-fauna of this bay, especially as it so freely communicated with the open sea

[1] *6th Ann. Report, S. F. B.*, p. 29.

beyond. The very indefiniteness of the bay made it in some respects a suitable contrast to the other areas.

The closure of the inshore waters of the Moray Frith, from Ord of Caithness to Kinnaird Head, was, however, substituted for Aberdeen Bay on the responsibility of the Board, and a series of stations for trawling marked off. As these had only been trawled once in 1887 no conclusions could be given.

Twelve stations were also marked off in the area of the Clyde—"six stations from the eastern side of the Mull of Cantyre to the entrance of Lock Ryan, three in Kilbrennan Sound, and three in the main water-way between Arran and the Ayrshire coast." A single series of trawlings were conducted in the spring of 1887, and it was concluded that "the flat fish of the Clyde are only about half as numerous as in the Forth at the same period of the year." It was further stated that this difference in the Friths was due to the prevention of trawling in the one and its prosecution in the other.

In referring to the numerical variation at different seasons and the relative proportion of small fishes captured, the reporters observe that in the Frith of Forth in June and July, 1886, the round fishes were more numerous than the flat, but that on all subsequent occasions the flat fishes were most numerous—"There can be little doubt that this was very largely owing to the suspension of trawling, which naturally affects the flat fish more than the round." Further they found that the adults of all the flat fishes were most numerous in June in the Forth and St Andrews Bay, with the exception of dabs in St Andrews Bay, which were rather more abundant in May. They therefore concluded "that there was a general movement of adult flat fish" into St Andrews Bay and the Forth at that time. On the other hand, the small flat fishes in the two localities were most abundant in August, with one exception, viz., plaice in St Andrews Bay, which were most numerous in October. They show, like the larger, a diminution in late autumn and in spring "probably from the same cause." The reporters did not find the same uniformity among the round fishes, and it is right to state that their deductions were somewhat premature.

They also made a comparison of the small (immature) fishes captured in the trawl of the "Garland," and by the liners at Buckhaven. Their statistics seem to be in need of explanation, but taking them at their own computation they consider that for every nine haddocks caught by the hook twelve are captured by the trawl. They did not add, however, that everything depends on the nature of the area.

An endeavour was made to utilize the statistics kept by about forty line-boats between Leith and Aberdeen for the purpose of determining whether any change had ensued since the closure. So far as can be ascertained from a perusal of the remarks on this subject no salient feature can be found, except that small haddocks formed a large proportion of the captures between July and December.

Further, a comparison is made of the amount of fishes exclusive of sprats, mackerel, and herrings landed respectively by the line-fishermen and beam-trawlers. In 1887 the average monthly amount of fishes captured by the liners was 3,500 tons, a far larger amount than by beam-trawlers. In the northern counties it was 1,925 tons, while that of the trawlers was 275 tons. If herrings, mackerel and sprats are included the average amount for the lines is 13,100 tons per month. In the same way the average monthly catch of flat fishes by liners was 200 tons, while the trawlers landed 300 tons.

The reporters in their summary state that the results of the "Garland's" investigations, as set forth in a special report to the Secretary for Scotland, "make it clear that the suspension of trawling has been followed by a great increase in the numbers of the fish within the protected areas both in the Frith of Forth and St Andrews Bay." "The increase has been shared by the round fish as well as by the flat; but, as was naturally to be expected, the augmentation has been proportionately far greater in the case of the less migratory flat fish." "In the Frith of Forth the more valuable of these, such as the lemon sole, have more than doubled in numbers since trawling was prohibited, and a similar result has occurred, and to a still greater extent, in St Andrews Bay, where the increase has been very marked." They also cite the case of the estuary of the Clyde as an instance of ground which "from the scientific evidence obtained, and from the testimony given on the spot," is naturally adapted for the support of multitudes of edible fishes. Yet fishes are much less plentiful there than on the east coast, since the trawlers have swept the waters, and have now left them as unremunerative. They also assert that a regular migration of flat fishes takes place during the winter, but do not enter into details, and state further, that small fishes are chiefly taken in inshore waters.

In St Andrews Bay in 1887 an increased number of hauls, viz., twenty-one, were made in the months of May, June, August, and October, and they yielded 2,649 saleable fishes, or an average of 126 per haul, a great advance on the previous year. Nearly one half (1,249) were plaice, and out of this number only three were large. Most were captured within the Bay, and in June a large number at Station IV., skirting the sandy beach. Dabs follow (838), and though generally distributed the greatest numbers occur on the inner Stations (I. and II.). Of this number 63 were large. With considerably smaller numbers, are haddocks (200), only 9 being large, flounders, gurnards, long-rough dabs (three being large), grey skate, whitings and lemon-dabs. Three soles and a cat-fish complete the list.

The unsaleable food-fishes reached the large number of 5,525, two forms, as before, being greatly in excess of the others and nearly equal, viz., plaice 1,786 and dabs 1,767. Then follow gurnards (956), which reached their maximum in June; small haddocks (810), which reached their maximum in October, the highest catch being on the outer Station (V.). That these were young fishes for the most part making for the inshore is probable, since only eleven were got in May, whereas in the subsequent months, June, August and October, the numbers are respectively 202, 152, 280. Lemon-dabs, long-rough dabs, grey skate, flounders, turbot and starry rays follow. Only four small cod, three whitings, one ling, and one sole occurred.

In connection with the increase, greater dexterity may have been acquired in the use of the trawl and, further, the incidence of the work in such favourable months as those mentioned,

The latter alone in such an exceptional year as 1887 seems to be sufficient to account for the increase, as will subsequently be shown. On the other hand, a very large number of plaice were removed from the bay by the lines of the fishermen, who, being undisturbed, kept relays of lines (often three sets) at work by day and night in the early part of the year. They observed, with truth, that already (March) they were obtaining better takes of flat fishes, and stated their belief that in a few years the inshore grounds would regain their former richness, thereby meaning that the larger fishes, which they affirmed they formerly caught, would again be abundant. It is doubtful, however, if due attention was given to the fact that they obtained the larger supplies by special exertions (even collecting lob-worms by torchlight), exertions which they had been prevented from employing, for at least twenty years, and then the maximum for each boat was two sets of lines.

In surveying the stations where the saleable fishes were captured, the inner (Station IV.) skirting the sandy beach, while furnishing only 24 plaice in May, gave no less than 581 in June, 55 in August and none in October. The unsaleable food-fishes were grouped differently, none occurring on Station IV. in May and June, while in August and October there were 236 and 757 respectively. The distribution of the dabs was more equable, but still with a tendency in the unsaleable to increase in June and August, whereas the saleable were most numerous in May. It is not easy to formulate the principle, if any, on which these variations rest. It might be said that in May the smaller plaice of the season had not had time to affect the captures, being too far inshore and too small—becoming greatly in evidence in June, August and October, but the irregularity of the figures forbid such conclusions. Nor does a consideration of the figures suggest that, as the reporters for the Fishery Board state, the great increase in 1887 is most marked on the inner stations, except in the case of dabs. The increase in the number of lemon-dabs in the outer Station (V.) is probably due to its being nearer their adult habitat. In regard to the other fishes, the increase in grey skate was marked, and saleable, and especially the unsaleable, haddocks

"Shelling" mussels.

To face p. 112]

were more numerous, as also were the whitings, long-rough dabs and turbot. The saleable gurnards, on the other hand, were considerably diminished, while the unsaleable had largely increased. The reporters do not take notice of the increase in the cod captured by hook in the bay, nor of the large numbers of plaice and dabs secured in the same manner, but such would have strengthened their position in regard to the results, and also somewhat altered certain views, *e.g.* the diminution of cod.

The uncertainty in fishing operations was demonstrated next year (1888) in which 25 hauls of the "Garland's" trawl from July to December (November alone being blank) produced the large number of 6,218 saleable fishes or 248 per haul, the sizes being now affixed, and only 595 or 23 per haul unsaleable, the total being 6,813, considerably less than the previous year, though the number of hauls was greater. As before, plaice were most numerous (2515 saleable and 135 unsaleable), dabs being next with 1937 saleable and 436 unsaleable. The saleable plaice increased steadily from July to October, viz. 479 in July, 518 in August, 638 in September, 645 in October, and suddenly fell to 21 in December. Only a single large plaice 26 inches long was obtained. On the other hand, dabs were irregular in this respect, viz. 335 in July, 787 in August, 685 in September, and 209 in October, the number falling in December to 21—the same as in the plaice. Large numbers were characteristic of good weather and the warm season, the numbers suddenly falling with the approach of winter and rough seas, when the trawl would work less efficiently[1]. A great increase appears in the saleable haddocks (863), which were mostly caught in July on the inner stations as they roamed inwards in their second year. No unsaleable haddocks were captured. The saleable gurnards had largely increased (503), but the unsaleable had as largely diminished (14 instead of 956), though what a saleable gurnard was may have given rise to doubt. The saleable skate had increased, but the unsaleable had diminished. The

[1] Dr Fulton in the *S. F. B. Report* (8*th*) notices that the largest proportion of immature to adult plaice occurred in St Andrews Bay in summer and autumn.

number of cod captured by the "Garland" was small (8), showing that her trawl was not adapted for this fish. Turbot had rather diminished, and there were no brill. Long-rough dabs and whitings had somewhat increased, but not much. If the months are taken separately, the totals are—for July 1834, August 1756, September 1517, October 986, December 126. These figures do not appear to give support to the gradual accumulation-theory, but rather lean to the notion of seasonal variations and effective working of the trawl. If the stations are taken in series, I. furnished 808 (average 178), II. 898 (average 243), III. 1919 (average 390), IV. 1623 (average 329), and V. 627 (average 221), that is to say, the centre of the bay was most prolific, the inner line along the sands being next. Former experience, however, showed that none of the lines were in the most productive region, which lay somewhat within Station IV., or "Scoonichill in a line with the steeples" as the fishermen have it, and in shallower water.

At every station in 1887 and 1888 an increase had taken place in the number of the food-fishes when contrasted with 1886, yet the irregularities were noteworthy. Thus at Station I. the numbers were more than doubled in 1887, but fell in 1888 to only 72 more (891) than what they were in 1886. At Station II. the increase was still greater in 1887, viz. 221 as against 1,636, falling in 1888 to 1,215. At Station III. along the middle of the bay the numbers were in 1886, 651, 1887, 1,107, and 1888, 1954, showing a steady increase throughout. Station IV. produced similar results, viz. 792, 1,484 and 1,647. The outer Station, V., again was irregular, viz. 505, 1,429, and 1,106. But it is well to remember that 4 additional hauls of the trawl were made in 1888, consequently the stations yielded averages during the three years as follows[1]:—Station I. 163, 479, 178; Station II. 73, 409, 243; Station III. 213, 276, 390; Station IV. 264, 571, 329; Station V. 168, 307, 221. These figures will show that the increase which marked 1887 was, with the exception of Station III., diminished in 1888. The diminution of fishes in the majority of the Stations in

[1] Calculated by dividing the total number of fishes (saleable and unsaleable) on each station by the number of hauls on that station.

1888 again demonstrates the uncertainty of fishing and of statistics, at least as regards preconceived views. The reporters observe, with truth, that the diminution in 1888 was chiefly due to the flat fishes, and it is found that the average catch per shot in 1887 was for plaice 144, and for dabs 124, whereas in 1888 the averages are 106 and 94 respectively. The meagre captures of dabs and plaice at the outer station (V.) in 1886 gave the reporters a basis for remarking that here round fishes had diminished but flat fishes had increased in 1888. The uncertainty of such observations, however, must always be borne in mind.

If we glance at two important fishes, viz. turbot and brill during the three years, and taking the saleable and unsaleable together, it is found that 13 turbot and 2 brill were got in 1887, 8 turbot in 1888 but no brill, while the first year (1886) only 2 turbot and 1 brill were procured. These small figures follow the rule of the larger, but give no certainty.

The influence of a month like December in reducing the percentage per haul of the fishes is clearly shown in 1888, for if we deduct the small number procured in this month (a total of only 157 in five hauls) from the result, and take the average of the rest it is found that the remaining 20 hauls give no less than 334 per haul, a result only a very little behind the condition in 1887 (389 per haul). The bearing of such facts will be clearly shown in subsequent pages.

This year the Board sanctioned the testing of the area in St Andrews Bay by a powerful steam-trawler[1] in February; but no systematic table of results is given. Five boxes of plaice, the largest specimen 17 inches and the average about 13 inches, and a few other fishes were obtained in one haul, and about two boxes of mixed fishes in a second haul of 4 hours. The lines taken in this case do not seem to have been productive. At least a small fishing-boat has not unfrequently brought to shore a larger amount as the produce of its lines.

In 1889 a considerable increase in the number of the hauls of the trawl took place, viz. nine more than in 1888, making a total of thirty-four. These gave 6,010 saleable fishes and 1020

[1] The "Southesk," belonging to Messrs Johnston, of Montrose.

unsaleable. The hauls were made in January, March, April, June, August, October and November, and thus both the colder and the warmer months were represented. The average per haul of saleable fishes was 176 and of unsaleable 30. This, therefore, when compared with the previous year, shows a considerable diminution (72 per haul) of saleable fishes, and a slight increase (7 per haul) of unsaleable. If the stations are taken in series, as before, Station I. has an average (saleable and unsaleable) per haul of 245—67 more than in the previous year. A diminution of 19 per haul takes place at Station II., and it is little more than half the catch of 1887. On Station III. a remarkable condition is found, the average being no less than 88 below what it was in 1886, not a third of what it was in 1888, and not half what it was in 1887. The captures on Stations IV. and V. are also much below what they were in 1888. Such variations appear to be connected with the season, the captures indeed forming a spindle from January to December. As Dr Fulton pointed out in the summary of the year, plaice had increased but dabs had diminished; thus in 1888 the average per 'haul' was for plaice 106 and dabs 94, whereas in 1889 the averages were 117 and 54 respectively. It is also observed "that at Stations I. and II. there was a considerable increase of flat fishes (almost entirely of plaice) and a large decrease at Stations III. and IV. (chiefly of dabs and plaice)." The figures in our tables show that in 1888 the total captures of plaice were as follows:—Station I. 284 or 56 per haul[1]; II. 506 or 101 per haul; III. 451 or 90 per haul; IV. 945 or 189 per haul, whereas in 1889 the figures were: Station I. 830 or 118 per haul; II. 736 or 105 per haul; III. 551 or 78 per haul; IV. 1561 or 223 per haul. For plaice, therefore, there is an increase on three stations. In regard to dabs in 1888, Station I. 406 or 81 per haul; II. 439 or 87 per haul; III. 612 or 102 per haul; IV. 450 or 90 per haul; and in 1889, I. 418 or 59 per haul; II. 535 or 76 per haul; III. 212 or 30 per haul; IV. 77 or 11 per haul. The increase of flat fishes, therefore, at Station I. was as 177 (1889)

[1] Saleable and unsaleable on each station by number of trawls on each station.

to 137, and at Station II. a diminution as 181 (1889) to 188—a comparatively limited increase. The decrease at Stations III. and IV. was respectively 84 and 45 on the total. It is interesting, however, to find that Stations III. and IV. united gave for the two years a considerably higher figure than I. and II., viz. 813 to 683.

This year the Herring Fishery Act came into operation, and the whole coast within the three-mile limit was closed.

During 1890 the hauls of the "Garland's" trawls were 30 in number, made during the months of March, May, August, October, and December. They produced 6,395 saleable fishes, or 213 per haul, and 2,273 (75 per haul) unsaleable, or a total of 8,668. There was therefore a decided increase on the previous year, both with regard to saleable and unsaleable fishes. This increase was chiefly during the warmer months, viz., May and August, and consisted for the most part of 3,472 small plaice from 7 to 11½ inches, and of 1,039 dabs of the same dimensions. On Station I. an increase of 9 fishes per haul occurred as compared with the previous year, while on Station II. there was a diminution of 3. An increase of 58 took place at Station III., while at Station IV. the large average of 721 was present—nearly double what it was in 1887. On the other hand, the diminution at Station V. was so great as to make the result about a third of what it was the previous year, and not half what it was in 1886. The uncertainty attending fishing operations was thus illustrated. There is a slight discrepancy, again, between these figures and the remark in the Report that "In this area there was a general increase in the total catch of fish at all stations except Station II., where there was a slight decrease."

Of the other features of the table the small number of the turbot (3), haddocks (45), lemon-dabs (9), long-rough dabs (22), and soles (1) is noteworthy, as indicating no tendency to increase or accumulate in the closed area. The reporter truly says that a special feature of this year was the augmentation of the whitings (388), which was due, he thought, to the vast shoal of young in the inshore waters in 1889. Such accumulations, however, take place frequently in their season, for though in

1889 only 43 whitings were captured, yet in 1888 there were 258. In regard to the gurnards, it may be observed that they were caught on almost all the stations chiefly in May and August—thinning off in March on the one hand and in October and December on the other.

Those who had faith in the gradual-accumulation-theory must have been perplexed in 1891, in which the number of hauls was 33, made in February, July, August, October, November and December. Of saleable fishes there were 3,668, of unsaleable 1,665, giving a total of 5,333--or 161 per haul, a result considerably less than any year since 1886, when the bay was closed. The average number of saleable fishes per haul was 111 and of unsaleable 50. At Station I. the average catch was the lowest recorded—less by 32 than in 1886. At Station II. the number was only about half that obtained during the three previous years and a third that of 1887. It is true it was much larger than that of 1886, but the number then was very low (73). At Station III. the number, like that of Station I., was the lowest yet obtained. At Station IV. the same result happened, the number being no less than 122 below that of 1886. By one of the usual variations the fishes obtained at Station V. exceeded those of all except 1887. The only round fishes in considerable numbers were gurnards, which, totally absent in February, occurred during the other months, though only at Station I. in December. The largest number were captured in November. The occurrence of 71 more or less adult herrings scattered over every station in February shows that this fish may spawn in the bay, and thus the hordes of young herrings in March will be more readily explained than by any other supposition. They were chiefly taken at Stations I. and V. near the southern entrance to the bay. The 20 cod were all large, and were got only in February and December—showing that during the warmer months they are rare. The numbers of the important flat fishes, such as turbot (1), soles (2), brill (0), lemon-dabs (4), demonstrated that as yet the closure had not altered their distribution. Lastly, the diminution in the number of plaice, viz., from 5,070 in 1890 to 1,598 in 1891, was probably due to causes altogether independent of their actual condition in the area. The weather, mode of working, state of the tide, time of day or other cause may have intervened. Moreover, it is sufficient to note that three winter months, viz., February, November and December, are included. Such results as the foregoing, along with those obtained in the Forth could have formed no reliable basis for further closures. They therefore were probably made on other grounds.

The diminution alluded to in 1891 continued to mark the captures in 1892, during which 40 hauls of the trawl were made in eight months of the year, viz., February, March, April, May, June, October, November and December. These hauls produced 2,680 saleable and 2,277 unsaleable, the average of saleable fishes being 67 per haul, and of unsaleable 56. This result, therefore, falls far short of what was found at the commencement of the operations in 1886, indeed, only a little more than half the number of saleable fishes, and 13 fewer unsaleable per haul were procured. On every station there was a reduction on the numbers of the previous year—which was also noted for its scanty captures. Moreover, on every station except one the average was less than in 1886, the exception being Station II.—where there were 4 more.

A new Station (VI.) was established in connection with the bay—in a line running obliquely in a north-easterly direction—from the Carr Rock to a point a little south of the Bell Rock. Four hauls of the trawl were made on this unenclosed area, viz., in April, October, November and December, the average catch of saleable and unsaleable fishes being 133; a comparatively moderate number.

In reviewing the kinds of fishes, the reporter in the Blue-book for the year points out the decrease in the flat fishes, but shows that the numbers of the cod and the whiting are greater than in 1891, but the numbers of the cod are trifling, viz., in 1891—20 cod, and in 1892—45; if we consider that no less than 44 hauls of the trawl were made, a far larger number, viz., 80, having been caught by a single "shot" of a fishing-boat in the bay. In the same way the whitings had increased from 114 in 1891 to 356 in 1892, and gurnards from 449 in 1891 to 1,096 in 1892. The numbers of the turbot, soles, brill and lemon-dabs showed no special feature, certainly no marked increase.

In connection with the diminution in the average catch of 1892, it has to be borne in mind that four cold months, viz., February, March, November and December are included. The average captures during these are much less than in the four warmer months of April, May, June and October.

Thirty hauls of the trawl were made in 1893, and the results were not more encouraging in regard to numbers than the previous year. If it were possible that the fishes gradually became from year to year more wary of a constant danger, as in the case of land-animals, such as grouse, for instance, the gradually diminishing totals might have an explanation. Yet though this may not have been the only factor, for the larger forms were also in all probability fewer, it is possible that it was one of them—both in the inshore and offshore waters. The total was no less than 839 under that of the previous year, grouped as 2713 saleable, and 1405 unsaleable fishes. The average number of saleable fishes in each haul was 90, and of unsaleable 46, the former average therefore being considerably over that of the previous year (which was 67), though both were comparatively small numbers. On the other hand, the average for the unsaleable was 10 below that in 1892. In regard to the various stations there was a continuance of the diminution on Station I. (125 to 102), an increase from 77 to 111 on Station II., a diminution on Station III. (from 152 to 115), a great increase on Station IV. (159 to 265), a diminution on Station V. (105 to 93), and a diminution on Station VI. (532 to 279).

In the Twelfth Annual Report (1893) there was said to be a slight increase in the general average of fishes captured this year, the increase being chiefly in flat fishes, and taking the two most important flat fishes, viz., plaice and dabs, for the numbers of the others are insignificant, during the three years 1891, 1892 and 1893 the following result appears:—

Average per Haul.

	Plaice	Dabs	Total
1891	48	86	134
1892	41	34	75
1893	56	27	83

The total per haul of these was therefore greatest in 1891, and that while there was a rise in plaice in 1893, there was a diminution in dabs. On the other hand the total captures of haddocks for the three years were respectively 56, 227 and 858, a progressively increasing number.

In the summary[1] it was stated that on the whole there was a considerable increase in flat fishes and a great increase in round fishes in the waters from which beam-trawling was prohibited. Unfortunately, when contrasted with the total averages for 1886 the result is a diminution of 5 in 1893, a fact which disposes of the theory of gradual accumulation, and at the same time shows the uncertainty ever attendant on fishing operations. Again, the great increase which was noticed in the haddocks off the east coast generally this year affected the bay so far as to raise the percentage of haddocks from 5 in 1892 to 26 in 1893.

The relation of the paucity of fishes with the preponderance of cold months was further illustrated this year, for four out of the six, viz. January, February, March, and November were more or less in this category. If, for instance, the totals (saleable and unsaleable) for these months are investigated it is found that the fishes caught in the months of June and July are about five times as numerous. This feature has, perhaps, not been sufficiently estimated by those dealing with the subject. Some explain it by asserting that there is a summer migration of flat fishes (*e.g.* plaice and dabs) to the shallower water; while others perhaps suppose that these fishes keep in the sand in cold weather. No observations made by the "Garland" clear up this point, a feature, however, familiar to fishermen.

[1] *12th Ann. Report*, iii. p. 32.

The total number of fishes captured in two hauls at Station VI. was 279, about half the number of the previous year, but the average was higher, viz. 139, or six more than in 1892. The average in both years was thus above the average of the closed area, which does not seem to point to the conclusion that the outer waters had suffered by non-protection.

During the year 1894 twenty hauls of the trawl were made in St Andrews Bay, viz. six in February, April, September and December respectively, each of the stations having been examined four times. The total number of fishes including the additional four hauls at Station VI. was 4238 or 2575 less than in 1888, with one haul more (or 25). The area, moreover, from which the fishes were drawn was considerably larger, for while there were only five stations in 1888, there were six in 1894. If the five regular stations only are dealt with, the total number was 3500, or 175 per haul, a considerable increase on the previous year. The average number of fishes per haul was for the saleable 138, and unsaleable 36, numbers which likewise contrast with those of 1888, the average in which was 248 for saleable, and 23 for unsaleable. The total number was under that of the previous year though differently apportioned, the average for the saleable being considerably larger, the unsaleable being 668 less. In every stage, therefore, the uncertainty attending these operations was demonstrated. Thus more fishes were captured on Stations I. and II. than in the three previous years. A slight increase on the previous year occurred at Station III., yet it did not bring it up to the captures in 1886. The numbers on Station IV. were only a little more than in 1893, viz. 287 to 265. On Station V. there was an increase of 63, while on Station VI. the average was 45 more than in the previous year, a fact not in consonance with the supposed depletion of the outer (unprotected) waters.

The average per haul, over the whole of the stations, for plaice was 48 and for dabs 64, a diminution of the former by 8 and an increase in the latter of 37, or more than double that of the previous year, the numbers of the latter fish raising the total considerably over that of 1893, viz. from 83 to 112. If the five enclosed stations alone are considered, the average for plaice is 59 and for dabs 24, an increase in the former case of 18 per haul and a decrease in the latter of 7 per haul. It is pointed out in the 13th Report[1] that there was an increase on the general average in the enclosed waters, and justly so, since the average per haul in 1893 was 148, whereas in 1894 it was 179 on the four areas. The reporter also observes that there was a considerable decrease of round fishes. Haddocks, however, show a slight increase from 26 to 29 per haul. The total cod caught on all the stations in 1894 was 17, whereas in 1893 it was 34; gurnards in 1893 were 13 per haul, in 1894 only 10. The numbers of the lemon-dab, turbot, brill and sole were insignificant, and, as previously, showed that the closure had no tendency to raise them.

A feature deserving special notice is the fact that the total of all kinds of fishes was the lowest in the series with the exception of 1886, with its 17 hauls. Two of the months were wintry, viz. February and December, while of the other two, April holds an intermediate position. Thus only September produced comparatively large numbers. A ship that could only work in good weather was, besides, unfitted for such investigations.

During 1895 the work of the "Garland" in St Andrews Bay was continued in March, April, June and October, and resulted in the capture of 4469 saleable fishes and 1802 unsaleable, the total of 6271[2] being the fifth highest during the 10 years, those of 1887, 1888, 1889 and 1890 exceeding it. But, as in other cases, this high average, if we may judge from the irregularity of the figures, seems to have no special significance further than that three of the months were mild, and usually characterised by high averages, while the third was less a winter month than for example January and December. The average per haul (313) was the second highest in the whole series of the decade, the exceptionally favourable conditions of 1887 alone enabling it to surpass it at the high figure of 408. On every station

[1] p. 19.

[2] Station VI. is dealt with separately.

a notable increase on the previous year was evident, and on Station IV. the number exceeded all except those of 1887 and 1890, the two highest in the series. Plaice showed a large increase per haul on the previous year, while dabs were more than doubled. It is quite legitimate to draw attention to the increase in such forms in St Andrews Bay in connection with the supposed influence of the closure in the Forth. There it is stated that an increase occurred in the protected area, and a diminution in the open area. But St Andrews Bay draws its supplies of eggs in the case of plaice from the open area, and should have suffered equally, which it did not. Lemon-dabs were four times as numerous, long-rough dabs were less (17 to 12, total numbers), while flounders had considerably increased on the preceding number (total 89 to 140). Of round fishes, cod were only 12 as compared with 17 in 1894. Haddocks were 15 per haul instead of 21, whitings were only 38 instead of 85, but gurnards showed the high average of 44 per haul, the second highest in the series, the highest being in 1887 with 49 per haul. As some of the numbers, *e.g.* cod, were small, these remarks are of limited value, but they show that the increase was chiefly in dabs and gurnards, fishes of secondary importance.

The three hauls on Station VI. produced 679 fishes or 226 per haul, the second highest number in the four years. The unprotected waters did not therefore give evidence of progressive deterioration from over-fishing.

In glancing at the results as shown in the averages of the various stations during the 10 years it would not seem that they point to a gradual accumulation of fishes due to the protection afforded by the closure. The higher numbers are succeeded by the smaller in such a way as to remove doubt on this point. Moreover, if they (higher figures) have a connection with anything it is with the warmer months of the year, that is, with the preponderance of these during the working period. This feature has been overlooked in the review of the 10 years' work[1], for instance, in contrasting the years 1887 and 1892 as illustrating the marked fluctua-

[1] *14th Report S. F. B.* p. 135.

tions during the period. In 1892, however, one half the hauls were in wintry months, viz., in February, March, November and December, while an intermediate month—April—also tended to lower the percentage. The comparison, therefore, with such a year as 1887, in which the hauls were all in productive months leads to erroneous conclusions. In dealing with St Andrews Bay, further, the remarkable activity of the local fishermen in scattering their lines throughout the area for the capture of plaice must have tended to interfere with the exact observations of the closure[1]. This point indeed was specially brought up at the Royal Commission in 1884 when the closure of the bay was suggested, but, as it would have been a hardship on the local fishermen, it was decided to exclude trawlers only. This factor therefore is one that should not be overlooked.

In the report, for the better illustration of the position, the first half has been contrasted with the last half of the period, viz., 1886–1890 with 1891–1895, and it was found that the average for the former period was 290 per haul, while in the latter period it was only 184 per haul, or a reduction of more than 100 fishes per haul. A basis is thus afforded for concluding that this progressive decrease, especially in flat fishes—so predominant in St Andrews Bay, is due to two causes, viz. general over-fishing in the open area, and, secondly, the destruction of too many of the spawning fishes on which the supply of floating eggs is dependent. Therefore it is proposed to extend the closure to the open area beyond the three-mile limit where such would include the "spawning areas" of the food-fishes.

But, before suggesting a step so important it would have been well to study the foundations on which the opinion rested. If a map (Table I.) is made of the months during which the hauls took place in the first period, viz., from 1886 to 1890, it will be found that the hauls are thickly dotted in the months of August and October, and have a preponderance in September. Briefly, two examinations of the areas took place in May in the first period (1886–90), whereas only one occurred in the second. Four examinations in August,

[1] *Vide* p. 112.

the most prolific month of all, augmented the averages of the first period; only one fell to the second. Two examinations in September were balanced by only one in the second period. Four examinations in October were followed in the second period by three. Now the averages per haul in all these months is high and rapidly affects the result; thus for May it is 253 per haul, August 515, September 362, and October 223. It is true that the examinations are balanced by three in each half decade in June, and that in July the latter (1891–95) has one period more (two examinations having been made in July, 1891), but the average for that month is not so high as some of those mentioned, for it is only 241[1].

On the other hand, the latter period (1891–95) is handicapped by frequent examinations in the colder months, which, while increasing the number of hauls, seriously affect the averages. Thus four examinations (20 hauls) in February, with an average of only 35, stand alone, for no examination occurred in this month in the first period. In March the examinations were equally balanced—three in each period. In April three examinations took place, as against one in the first period, the average for this month being 122; moreover there were 15 hauls in the three examinations, as against only five in the first period. When we come to November and December a similar condition is found. In the second period, three hauls in each are balanced by only two in the first period, so that their comparatively low averages (191 for November and 39 for December) affect the second period considerably, especially as there were 15 hauls in November and 13 in December, as against 10 in each of these months in the first period. In the first period 25 examinations were made; in the second 29, yet though the latter has the preponderance in number, the former has the larger proportion during the warmer months (which possess a high average), viz. as 9 to 13.

This important point is further brought out by calculating the number of hauls made during the six winter months, viz. January, February, March, April, November and December in each period. In these months, during the first period (1886–

[1] Food-fishes alone are included in these averages.

1890) there were but 44 hauls, whereas in the second period (1891–1895) there were nearly double the number, viz. 83. The same result is apparent by a comparison of the hauls during the warmer months of May, June, July, August, September and October, for out of a total of 126 hauls in the first period (1886–1890) no less than 82 occurred in these months, whereas in the second period (1891–1895) out of a total of 143 hauls only 60 belong to these months. In other words, there was a difference of 38 hauls in favour of the warmer months in the first period, and an excess of 23 in favour of the colder months in the second period (1891–1895). Between the hauls in the warmer months of the first period and those of the second period, there was thus a preponderance of 22 hauls to the former period (1886–1890).

St Andrews Bay.

1st Quinquennial period	Warmer	Colder		Difference
	May to Oct.	Jan. Feb. Mar. Apl. Nov. Dec.	Total	
No. of hauls	82	44	126	38 + in favour of warm
2nd Quinquennial period				
No. of hauls	60	83	143	+ 23 in favour of cold

From most points of view, therefore, the doubtful nature of the deduction on which further closures rested is sufficiently obvious[1]. The weight to be placed, therefore, in the Summary[2] on the statement that there was "a general falling-off in the abundance of flat fishes" "especially in the closed waters of St Andrews Bay" in the second period is evident. This contrast between the two quinquennial periods, with the table, has been quoted from the Fishery Board Report by Mons. G.

[1] The differences between the captures in the warmer months (June and July) and those of the colder were indeed adverted to in the reviews of the first results of the "Garland's" work by the late Sir James Gibson Maitland and Prof. Ewart, *5th Ann. Report S. F. B.* pp. 57 & 58.

[2] *14th Ann. Report S. F. B.* p. 137.

Roché, French Inspector in Chief of Fisheries, as if it were perfectly reliable[1].

The accompanying tables (I. to VII. *et seq.*) will demonstrate that the basis on which the foregoing criticism rests is not imaginary. The percentages increase (with a slight check in July) to August, and then diminish to December, just as the spindles formed by the animals in the pelagic fauna do, and as the larval and post-larval fishes do from January to December[2]. The argument in the Report of the Fishery Board[3], therefore, for further closure so as to control the "spawning areas" rests on no satisfactory basis in this connection—whatever support it may have from general over-fishing. The conditions of the two half-periods were wholly divergent, indeed, it might well have happened that the differences were more pronounced than those in the Blue-book. In the earlier Reports of the Board such variations would have been attributed to migrations, but there is no reason to think these are of any moment in this case. Fishes become more lethargic in cold weather, more lively in warm weather, and hence more readily come in the way of capture in the latter than in the former. Consequently, the statistical analysis based on the half-periods just mentioned fails to be reliable, and does not prove "that there has been a diminution of the more important flat fishes in the closed waters, instead of an increase, as was anticipated; and that this may probably be traced to the influence of beam-trawling in the open waters where the fishes spawn[4]."

The reporter, it is true, states that the same conclusions

[1] *La culture des mers*, 1898, p. 86, &c.

[2] *Vide Life Histories of the Food-Fishes*, McIntosh and Masterman, pp. 36—56.

[3] *14th Report*, iii. p. 148.

[4] One critic somewhat facetiously accuses the author of being responsible for this statement in the Fishery Blue-book of the year (1893). He doubtless reserved his opinion on this and other questions—the materials for decision in which did not appear to him to be sufficiently complete, or were unsatisfactory. As he had not then gone fully into all the details (which indeed were not then in existence) he deemed it right to fall in with his five experienced and valued colleagues. His critic perhaps has not observed that three other members, not unknown in various connexions with line-fishermen, dissented from the views he deems so important.

cannot be drawn "with regard to round fishes, which are more numerous and migratory." But if his conclusions are based on his half-periods, experience would show that it is during the colder months of the year that cod, and haddocks, and whitings most abound in St Andrews Bay, yet the uncertainty of fishing-operations is again demonstrated by the fact that while the cod and whiting were more numerous in the latter period, the haddocks and gurnards were in the majority in the former period. The highest average for haddocks took place in 1887, while the lowest years were 1889 and 1890.

The opinion expressed in regard to the cod in St Andrews Bay in 1883 and 1884 was that it had almost disappeared, but the examination during the latter year both in the bay and off the rocks showed that this was at variance with facts. Indeed, no sooner were the fishermen free to leave their lines baited with anemones than cod were found in considerable numbers. A single boat would bring from 30 to 80 cod as its share in a bay that had been "swept barren by trawlers." It was clear that only the right season and the right bait were necessary to secure fine cod as before. Year by year the same experience has been met with. Lately (1898) even so limited an apparatus as a salmon stake-net in June has captured at a single tide upwards of 6 dozen cod from 14 to 27 inches, showing that notwithstanding all the pessimistic views of those engaged in the pursuit, the nomad shoals of the round fishes still assert themselves. Besides, the yearly shoals of tiny young cod in the bay in June show, by their unfailing regularity and immense multitudes, that the methods of nature in the open ocean are for the most part beyond the influence of man, and go on as before—unmindful alike of the fears of the timorous, and the misapprehensions of those imperfectly acquainted with the subject.

If the averages on the stations in the closed area are considered, the same result is apparent. No accumulation and no encouraging increase are evident, but the lower and higher numbers succeed each other in a way explicable only by the irregularities and uncertainties of such operations. Thus the average 175 in 1886, just after powerful steam-vessels and fleets of local trawlers had swept the bay, was succeeded, after an interval of four years by even smaller numbers for three consecutive years, and a fourth year by the same figure. Further, as if to dispose of the view that the general over-fishing, or the detrimental influence of free fishing, in the open waters, had slowly sapped the supply, the last year of the "Garland's" work had a very high average. The highest of all the averages was in 1887—the year after the closure, the next in 1895. Again, the comparatively high average of 288 in 1890 was followed by low averages for three years. The reporter in his summary[1] of the results of the "Garland's" work considers that "the immediate consequence of the cessation of trawling appears to have been an increase in the abundance of flat fishes within the enclosed areas;" and, further, "the fact that this increase was not only not maintained, but that a progressive decrease in plaice and lemon soles occurred subsequently, indicates another influence, namely, excessive trawling on the offshore grounds where these fishes spawn."

It should be borne in mind that the work in 1887 was done during productive months, viz., in May, June, August and October, a fact which alone would considerably add to the percentage, since not only flat fishes, but gurnards and other forms would swell the lists. No winter months increased the number of hauls, while it diminished the percentage of captures. Nowhere in the tables is such a combination of favourable conditions to be found, and consequently in no year is the average so high.

Much has been said about the disappearance of the older and larger plaice which were so common about the period when trawling began, but surely no other result could have been expected. Any method of fishing persistently and effectively carried out on an area where flat fishes occur has a tendency to remove the larger examples of such fishes as plaice, turbot, halibut and other forms, but this does not mean that the race is extinguished. The "practical man" forgets that there are other sizes of plaice, that are, for instance, just as useful to mankind as the 20-inch plaice. He forgets that year by year,

[1] *14th Report*, iii. p. 148.

irrespective of all melancholy forebodings, myriads of young plaice people the shallow waters of the sandy bays all round our shores, and afford the best refutation of his views, and a proof, beyond dispute, of the recuperative powers of nature. For, just as surely as these swarms of young plaice annually appear, just as surely are the ranks of the sizes above them recruited. Does the "practical man" find that the dabs and long-rough dabs have been similarly dealt with by the trawl (which makes no discrimination in regard to species)? Large plaice and large turbot are certainly conspicuous examples of forms that are affected by constant and effective methods of fishing, but it is long since their extinction was prognosticated, and, so far as facts go, their continuance is as yet in no danger.

The opinion expressed in the summary that the dabs and long-rough dabs by their increase in the closed waters had to some extent supplanted the plaice and the lemon-dabs seems to be in want of further proof. The view that it is the trawl which has caused an undue slaughter of the latter (before maturity), while the two former species escape as mature individuals through the meshes, will not explain all the facts for St Andrews Bay. Moreover, the reporter has forgotten that the flounder is in the same position, its numbers having been reduced by about a fourth in the same period (latter half), yet it cannot be said that this species depends on the outer water for its eggs and young[1]. A better case, as regards diminution during the second period, could be made out for the turbot, since in the first period 34 were captured, and in the latter only 18. Ripe turbot as a rule are offshore fishes, and their destruction, therefore, was as probable as that of the spawning plaice. Yet he would be a bold man who would take up such a position, especially if familiar with St Andrews Bay, in which the young turbot have always occurred. Similar remarks apply to the sole, the first period having double the number of the second. But if the "Garland" had had a ground-rope on her

[1] It has been stated in the Fishery Reports that flounders have increased because they spawn inshore, while lemon-dabs have diminished since they spawn offshore, and their young are not confined to the territorial waters. A careful study of both cases does not substantiate this view for either.

trawl of the same kind as that formerly used by the native fishermen the result might have been different. Few would be prepared to state that because, even with a blank year, the thornbacks were more numerous in the first period (by 107 to 97) they had been affected by any method of fishing, just as they would hesitate to put weight on the fact that the small numbers of brill were increased fourfold and the larger numbers of starry rays doubled by accumulation in the latter period (*vide* Tables V.—VII.). The uncertainties and irregularities connected with fishing operations are as distinctly seen in the fishes that are less common in the bay, such as the sole, sandy ray, conger, poor cod and wolf-fish, as in the more abundant forms, some occurring in greater numbers in the first period and others in the second. The greatest caution is therefore necessary in drawing conclusions from such data. If they show anything more than another it is that the closure of such areas has no appreciable influence on the increase or diminution of the fishes within their own boundaries or in the neighbouring waters. The problem is too vast to be solved by such pigmy measures. The closure—to the three-mile limit—round the shores of Britain seems to the uninitiated a great step for the protection of the sea-fishes against trawling, but it is really quite insignificant, in view of the vast extent and unbounded resources of the ocean, or in its effects on the complex chain of circumstances resulting in the permanent abundance of these fishes.

St Andrews from the Pier. The Castle is on the extreme right, and St Leonard's School on the sky-line on the extreme left.

To face p. 132]

CHAPTER IV.

Scientific Investigations in the Frith of Forth. 1886—1896.

Tables VIII. to XII.

The area of the Forth, which comprises 250 nautical square miles, differs in a marked manner from the open sandy bay of St Andrews, both in regard to the nature of the bottom and the depth of water. These distinctions are clearly set forth in the Fifth Annual Report of the Fishery Board[1] in the description of the nine trawling stations of this area, the salient features in contrast with St Andrews being the frequent occurrence of mud, stones and shells as well as sand, with such forms as oysters, mussels, clams and Norway-lobsters. The depth, moreover, ranges from 10 to 20 fathoms. Though much marine zoological work has been done for generations in the Forth, the precise condition of its fish-fauna just before the closure was less accurately known than that of St Andrews Bay, mainly because the trawling ships chiefly worked beyond its limits. It is true that the otter-trawl was used in 1858, and probably earlier, on the sandy flats, mainly for scientific purposes, but no accurate records of the captures are available for comparison. Accordingly, the first season's work of the "Garland" (in 1886), with such supplementary evidence of a reliable kind as exists, forms the chief guide in dealing with the area.

The first season's work in the Forth commenced in June,

[1] pp. 52—54.

1886, and was continued in July, September, October and November. Only in June, however, were the stations systematically gone over (with the exception of VIII.). Twenty-seven hauls of the trawl produced a total of 5,896 fishes, of which 4,649 were saleable and 1,247 unsaleable. Of the saleable fishes the majority were haddocks, no less than 2,273 having been captured, or an average of about 84 per haul. The largest captures were in July, *e.g.* 357 and 382 respectively occurring at Stations III. and VII. The average number per haul in July was 297, the next (78) being in June. The other three months were much less productive in haddocks. The deeper waters of the Forth probably prove more attractive to the haddock than the shallow waters of St Andrews Bay. The next were plaice—1,056, or 39 per haul, most having been procured in September, viz., an average of 80 per haul; dabs 791 or 29 per haul; lemon-dabs (total) 445; long-rough dabs 419; whiting 253; gurnards 279 or 10 per haul; cod 241. In contrasting these captures with those in St Andrews Bay the great proportion of haddocks is a prominent feature, about nine times as many being present in the Forth. Cod and whitings are also in larger numbers. Gurnards, on the other hand, were much more common, and plaice and dabs were twice as numerous in St Andrews Bay. Lemon-dabs and long-rough dabs were more abundant in the Forth, while flounders were more common in St Andrews. As already mentioned, the deeper water and more varied bottom, together with the periodical abundance of herrings probably proved a greater attraction to both cod and haddock. The absence of thornback for the first year in the returns of the "Garland" in St Andrews Bay leaves some hesitation about a comparison of the skate of the two areas, but on the whole they correspond, with the exception of the sandy ray, which occurs only in the area of St Andrews, a feature, however, of no moment. If we consider the months, it is found that July with its 4 hauls produced 2,223 food-fishes or 555 per haul, June with 8 hauls gave 1,313 or 164 per haul, September with 5 hauls 1,171 or 236 per haul, October with 4 hauls 576 or 144 per haul, and November with 6 hauls 613 or 102 per haul.

The average number of saleable fishes per haul was 172, of unsaleable 46, giving a total of 218, a number considerably higher than that of St Andrews Bay, a fact which traverses the suppositions so common about this period.

The work of the "Garland" in the Forth, therefore, showed that food-fishes occurred in considerable numbers throughout the area, especially when the inefficient means of capture is borne in mind. A well-equipped trawler pursuing its work for commercial purposes would have shown a different return, for it would have been free to select its ground according to weather, tide and season.

The highest average (423 per haul) occurs on Station III. from the centre of the Forth towards the southern shore and in 8—10 fathoms east of Inchkeith; the next (308) is at Station II. from Dysart to Leven in 9 to 12 fathoms; the third (211) on Station I. in the centre of the estuary and in 10 to 18 fathoms; the fourth (204) on Station IV. off the southern shore between Fisherow and Gullane Ness; the fifth comprises two stations, VI. and VII., which have nearly the same average, viz. 197 and 196, yet they lie on opposite sides of the Forth, the former being the short station (Fluke-hole, 13 to 14 fathoms) opposite St Monance and Pittenween, and the latter along the south shore westward of the Bass. The seventh place with an average of 169 is held by Station V. running in the centre westwards from the Isle of May and with a depth of 13 to 14 fathoms. These were all within the enclosed area. The others, viz. VIII. and IX. were outside the limit and with a depth of 18 to 20 fathoms, the former having an average of 114 per haul, and the latter only 46. It may be doubted, however, whether the trawl of the "Garland" was always effective in the deeper parts.

In 1887 the work in the Forth was carried on in the months of June, August, and September—all productive months. It forms a useful comparison with that in St Andrews Bay which was spread over four months, viz., May, June, August and October, also all productive months (*i.e.* with a high average). In neither area therefore did a winter month reduce the average. Here as well as in St Andrews Bay the increase in the fishes was noteworthy: the total was nearly doubled, though, just as in the case of St Andrews, it was by the great abundance of the unsaleable, indeed, the saleable fishes were no less than 1,392 below what they were the previous year, a fact which is important. With 3 hauls less than in 1886, viz. 24, a total of 9,094 or 3,257 saleable and 5,837 unsaleable fishes were obtained. The most abundant fish was the haddock, which formed about a third of the entire captures, viz. 3,101, or an average of 129 per haul. These fishes were therefore 45 per haul over those of the preceding season. The highest figures (saleable and unsaleable inclusive) occurred in August, apparently when the young haddocks were seeking the inshore waters from their birthplaces in the offshore, for 1,007 unsaleable forms and 100 saleable were then captured, the next in order with 1,012 being September, while June had 982. Plaice, as in the preceding year, followed with a total of 1753 or an average of 73 per haul, nearly double that of 1886, dabs with a total of 1157 or 48 per haul, lemon-dabs 1,129 or 47 per haul, long-rough dabs 682, whiting 383, gurnards 377 or 15 per haul. The area showed a considerable number of grey skate, flapper-skate and herrings (32), and a single mackerel, sole, catfish and bib were obtained.

As in the previous year, it was the presence of large numbers of small haddocks that chiefly raised the average of the captures to 378 per haul, just as in St Andrews Bay it was the increased number of plaice that raised its average to 341. The one seemed to be a harbour for the round fishes, the other for the flat fishes. The increase, however, was much greater in the case of the plaice.

The highest average occurred in August, for 7 hauls produced 3,004 fishes or 429 per haul; the next was June, 9 hauls with 3,317 fishes or 360 per haul, and lastly, September with 8 hauls gave 2,713 or 339 per haul. Though these were considerably higher than in the same months of the previous year, yet, in July 1886, the average per haul was still higher, viz., 552. Such are the uncertainties ever recurring in these pursuits, even with all the parallel advantages of favourable months.

The highest average is at Station II. off the Fife shore—

from Dysart to Leven, viz. 616 per haul, the highest obtained in the whole series. This station has both a considerable variety and abundance of fishes. The next is Station III., with 589 or 165 above its number last year—when it headed the list. The third is Station I. with 566 per haul. Then follow V., VII., IX., IV., VIII., all with a higher average than in 1886, while, on the other hand, VI., the short station opposite St Monance and Pittenween, had 42 per haul less than in 1886.

No subsequent year was on the exceptional footing of this year with its three productive months. It is interesting, moreover, to find that the average of the saleable fishes was exceeded by the previous year and by several subsequent years. The unsaleable and the total were unique (highest) in the series.

So far as can be ascertained little is to be gained by a comparison of the various kinds of fishes obtained in the outer and in the inner stations[1]. The broad fact remains that this year (1887) the average of the captures was very high, but, as the work was carried on only in three very productive months in the Forth, its bearing on the supposed increase by accumulation and non-interference is questionable. It might be said that the large number of immature fishes captured was a proof that the closure had at least preserved the young forms wafted into the Forth either as eggs, larval, post-larval or very young fishes. For the most part, they had complete immunity (by their size) from any kind of fishing apparatus outside the limit, and once inside the limit they were safe. Unfortunately, however, every subsequent year was in the same condition, and yet no steady and permanent increase took place in the fishes of the waters of the Forth. It cannot be said that eggs and larval fishes were rare beyond the limit. Floating eggs in great numbers near the surface, and countless swarms of young fishes in the deeper water occurred in the outer area, as in all probability they had occurred for ages. The eggs and pelagic young were also present around and inside the Island of May. Materials for an increase were thus in abundance.

[1] *Vide Sixth Annual Report, S. F. B.*, pp. 28, 29.

In dealing with the work of this and the preceding year in the Forth no support can be given to the conclusion indicated on pp. 108 and 109 "that the suspension of trawling had been followed by a great increase in the numbers of the fish in the Frith of Forth."

In 1888 the number of the hauls was greatly increased, viz. to 50, and they extended over a period of six months, two of which—November and December—were winter-months. These produced 9,952 fishes, or only about 800 more than the 24 hauls did during the warm months of 1887. Of this number the saleable fishes made 8,928, or more than twice the number of last year (leaving out of consideration for the moment the greatly increased number of hauls), and the unsaleable 1,024. The sizes were now affixed, so that comparisons with the previous year need not be made on this head. Haddocks held the first position, as formerly, with total of 2,225 or 44 per haul; only 14 were large, the majority being medium (1512) and small (692), the unsaleable amounting only to 7. The average was thus about a third that of last year. The highest number occurs in August (693), the next in September, then October, June and December, the lowest being November with only 105. The increase of the haddocks in December over those in November was probably due to the presence of herring in the Forth. Plaice again follow with a total of 2,028 all saleable, or an average of 40 per haul, considerably under the previous year. It is interesting to notice that the plaice follow a similar relative increase to that of the haddocks from June to October, but that a sudden and great fall takes place in November and December—out of all proportion to that in the haddock, the number in December being only $\frac{1}{11}$th of that in June, whereas in the haddock the captures in December were only 17 less than in June and in the case of the whiting 19 more than in June. It is the remarkably low averages in the two winter months in such as the plaice which accentuate the difference between 1887 and 1888. Whitings come next plaice with a total of 1,378 (only 30 of which were unsaleable) or an average per haul of 25, or 10 over the average per haul of last year. They followed the haddocks closely in regard to the proportions during the various months, though of course in diminished numbers. Next were dabs at 1,244, or 24 per haul —exactly half the number per haul when contrasted with the previous year. They followed the plaice in regard to the months, though the highest number occurred in September instead of in August. The fall was less striking in November and December, but it was likewise present. A large proportion of the dabs were unsaleable. The total number of lemon-dabs was 888 or 17 per haul—no less than 30 per haul below 1887; long-rough dabs had a total of 688. Cod and gurnards were fewer than the previous year.

In contrast with 1887, the stations throughout showed a reduced number of food-fishes in the average per haul, the first and third having only about half the number. The first two had about the same average as in 1886, while the third had little more than half that of 1886. The fourth and fifth were not very different from what they were in 1886. The sixth was less by more than half, the seventh, eighth and ninth had considerably increased.

This year eight hauls by an ordinary steam-trawler[1] were made by the Board in the Forth at the beginning of February as a check on the work of the "Garland," but no systematic treatment of the results is given in the Blue-book. It is simply stated that the main results corresponded with those obtained by the "Garland." It is unfortunate that an accurate table of each of these hauls was not preserved.

The lesson learned from this year (1888) was not a striking one, but it brought out the uncertainties of fishing operations (mainly perhaps due, in this case to stormy weather), and showed the greater variety in the commercial captures of the Forth as contrasted with St Andrews Bay. The various kinds of skate were scattered generally over the area, and such forms as the cat-fish (wolf-fish), green cod, herring, witch (pole-dab) and pollack, were perhaps more characteristic of the Forth. The conger, ling, sprat and halibut were sparse in both, while the haddock, whiting and cod were in greater numbers in the Forth.

Further increase of the number of the hauls took place in

[1] The "Ocean Rover" belonging to Mr Gunn, of Leith.

1889, viz. to 90, with the result that a total of 13,768 fishes or 152 per haul were obtained. Of these 11,981 or 133 per haul were saleable and 1787 or 19 per haul unsaleable, a reduction in every case on the previous year. Thus the total average is less by 46, the saleable by 45, and the unsaleable by 1. The comparatively small average number (15) of unsaleable fishes is noteworthy—since it is the lowest in the series. A stricter method of computation both the preceding year and this may have had some connection with the sudden fall in this column. At least four winter months were included in the working period, viz. January, February, March and November, besides April, which has not a high average. The more productive months were May, June, July, August and October. The absence of such a month as September makes a considerable difference, especially when the balance lies between five months with rather a low average, and five with a high average.

The uncertainty of striking a shoal of round fishes was illustrated in the case of the haddocks, the totals of which (1445) were less by 780 than that procured by only 50 hauls in 1888. Instead of 44 per haul their average was now only 16. They reached their maximum in August (33 per haul) and their minimum in the warmer months from April to July. The increase in August was probably due to the immigration of the summer haddocks of the fishermen, though other causes may have intervened. A similar reduction occurred in whitings, the total for which (1358) was 20 below that of the previous year with its 50 hauls. They also reached their maximum in August (32 per haul), their next highest periods being November, October and July; the numbers during the rest of the months being considerably less. The highest total was that of plaice, 3833 or 42 per haul, a somewhat higher average (by 2) than the previous year. They also reached their maximum in August with an average per haul of 57, June and October following. The minimum was in January (11 per haul) and the next lowest in May (16 per haul). Dabs had a total of 1930 or 21 per haul, three less than the previous year. They likewise reached their maximum in August and their minimum in January. The average per haul in the lemon-dabs was below (by 3) that of 1888. Cod and gurnards remained about the same in each year, as did others. The increase in the number of the "witches" in this and the previous year is a fact of interest.

All the Stations except IV., VI. and IX. showed a reduction in the average of the food-fishes captured. Station IV. had only an increase of 6, yet this was the highest average (272) of the year. The increase per haul at Station VI. was 31, and at Station IX. only 8.

In contrasting the diminished averages of 1889 with the previous year it must not be forgotten that the former has the disadvantage of including five winter months in its working-period, whereas the latter has but two. Under these circumstances it is perhaps remarkable that the reduction was not greater. Yet such does not indicate that a serious diminution of the fishes of the area had taken place; but is the effect of a well-known seasonal variation.

A considerable increase took place in the captures of 1890, seventy-nine hauls producing a total of 18,270 fishes or an average of 231 per haul. Of these 14,953 or an average of 189 per haul were saleable, and 3317 or an average of 42 unsaleable. The work giving this result was carried on during 10 months, five of which were winter months, viz. February, March, April, November and December, and five warmer or productive months, viz. May, July, August, September and October. While the three first mentioned winter months, however, produced respectively 501, 720 and 891 fishes of all kinds, the two last months of the year gave respectively for seven hauls 1970 and for nine hauls 1753 fishes, a very considerable addition to the totals. The captures in the warmer months were as follow:—May with 9 hauls 1479, July with 7 hauls 2432, August with 6 hauls 2284, September with 5 hauls 2067, and October with 9 hauls 4173. It is clear that if July, August and September had had as many hauls as October the result would have been very different, for these are productive months. In an inquiry like the present, therefore, it is important to have the data equivalent throughout, or at any rate to be cautious in drawing deductions from results obtained under different conditions.

The adding of new stations during the progress of the inquiry was also perhaps unnecessary, and thus for example Station X., which was arranged for in 1892, has simply been struck out of the reckoning.

The comparison of the closed with the open area, that is, of the stations beyond the limit with those within it has always appeared to produce slender results, and accordingly it has not been deemed necessary to allude to it.

The foremost place was held by the whiting with a total of 5,545 or 70 per haul, of which 5,093 were saleable and 452 unsaleable. Large numbers of young whitings had been observed in the Forth and neighbourhood by the "Garland" in the autumn of 1889. Dr Fulton, indeed, computed the immense shoal in territorial waters to reach at least 230,000,000 individuals between 3 and 5 inches in length. Nor is this condition to be considered as exceptional. The young whitings come inshore from the more distant grounds where the majority pass their early stages in similar shoals annually in autumn, but they are not always discovered, and probably may often be much more scattered. Very considerable captures of saleable whitings were also made in 1884 south of the estuary of the Forth, no less than 1,522 being brought up in a single haul of the trawl on the 7th February, 1884[1], so that the general diminution of this species is by no means proved. Trawl-fishermen consider, moreover, that the whiting is a species less apt to come in the way of the net than such as the haddock, since it often feeds higher in the water, a fact borne out by its active habits and piscivorous tastes. Besides, capture by hook or by trawl sometimes gives but an indifferent view of the number of fishes on an area. Of the foregoing total of 5,545 whitings no less than 3,095 were between 7 and 9 inches in length—captured in the following months:—

No. of Hauls			No. of Hauls		
9	February	25	6	August	182
9	March	38	5	September	608
9	April	133	9	October	1494
9	May	174	7	November	449
7	July	321	9	December	480

[1] *Report Royal Commission on Trawling*, p. 389.

Those under 7 inches (452 in number) were procured:—

No. of Hauls			No. of Hauls		
9	February	23	6	August	45
9	March	39	5	September	10
9	April	86	9	October	31
9	May	76	7	November	51
7	July	33	9	December	58

The captures of the larger series are sufficiently varied, but it is clear that there was a steady increase till October, when the maximum occurred. It might be supposed that the great increase in October was due to the growing fishes seeking the bottom after certain kinds of food. On the other hand, a glance at the table shows a tendency to a marked increase in the outer Stations VIII. and IX. in September, and then, next month, these and the stations in line with them, viz. VII. and III., were all accentuated, as if a shoal had about that period sought the inshore ramifications. According to present notions those between 7 and 9 inches would be in their second year in October, so that there may be a tendency shorewards also in their second autumn, probably in pursuit of a favourite food. The irregularity of growth may account for the prevalence of small whitings throughout the year, as shown in the list of those under 7 inches, though in the later months the most advanced young of the season may be included.

The haddock succeeds the whiting with a total of 2,474 or 31 per haul, nearly double what it was the previous year, the saleable being 2,291 or 29 per haul, and the unsaleable 183 or 2 per haul. Of the foregoing, 1,445 were between 8 and 9 inches and 183 under 8 inches. They were procured as follows:—

No. of Hauls		Under 8 in.	Between 8 & 9 in.	Between 10-14 in.	Large 15-26 in.
9	February	22		12	12
9	March	34		11	5
9	April	19	13	16	3
9	May	59	29	21	13
7	July		675	37	10
6	August	4	244	8	8
5	September	5	304	304	10
9	October	22	146	258	7
7	November	11	6	38	5
9	December	7	28	56	12

The maximum thus occurred in July, three months earlier than in the whiting, and the Stations V., VI. and VII. at the entrance to the Forth seemed to have been most productive. It is possible that the series under 8 inches beginning in August are those of the year, since we know with certainty that some reach 7½ inches in November. Out of a total of 85 large haddocks captured this season, 52 were obtained subsequent to the spawning-period, viz., from July to December. This proves that the 33 captured earlier in the season did not deplete the waters of the area of spawning fishes. It is probable indeed that some at least of the 33 had also shed part of their eggs, and as the adult may contain from 170,000 to 2,000,000 eggs, the number may have been considerable. But no account has been taken of the 761 between 10 and 14 inches, a proportion of which, especially amongst the males, between 12 and 14 inches had arrived at maturity. Besides those captured by the "Garland" or other means in the Forth, a large reserve-margin must have escaped. Thus, in whatever way we look at the question, there is no reason to suppose that we have reached a dangerous condition in regard to our marine fisheries—as a certain class maintain. A little more trust in Nature, and less in "friends," or in help from the Government, would not be for the disadvantage of the department.

Plaice come next in order with a total of 2,437 or 30 per haul, a result below what it was in 1886 by 9, but the latter period differed as regards the proportion of productive months. 2,422 were saleable and only 15 unsaleable, so that practically the plaice were nearly all marketable. In contrasting the captures of this fish in the Forth with those in St Andrews Bay in March, May, August, October and December of the same year, it is found that the following condition, as shown in the table, occurs:—

The sailing-boat "Dalhousie" of the Gatty Marine Laboratory with her "crew" on board.

To face p. 144]

1890.

FORTH.

Plaice, total 2437.

No. of Hauls		19-26 in.	12-18 in.	7-11 in.	Under 7 in.	Total
9	February	1	140	46	2	189
9	March	—	76	57	2	135
9	April	2	84	40	1	127
9	May	5	129	110	3	247
7	July	2	156	181	1	340
6	August	11	283	162	5	461
5	September	—	111	10	—	121
9	October	5	248	199	1	453
7	November	2	235	38	—	275
9	December	2	76	11	—	89
79		30	1538	854	15	

ST ANDREWS BAY.

Plaice, total 5070.

No. of Hauls		19-26 in.	12-18 in.	7-11 in.	5-6½ in.	Total
10	March	1	329	464	6	800
5	May	1	208	427	3	639
5	August	—	197	2464	725	3386
5	October	—	114	116	3	233
5	December	—	11	1		12
30		2	859	3472	737	

While 79 hauls of the trawl produce but 2,437 plaice of all kinds in the extensive area of the Forth, 30 hauls in St Andrews Bay give more than double, viz., 5,070. The apportionment of the various sizes of plaice is equally divergent. Thus of large plaice 19—26 inches in length, the breeding plaice of the areas, one occurs in every two or three hauls in the Forth with its deeper waters, and only one in fifteen hauls in St Andrews Bay. Of plaice from 12—18 inches, a very valuable marketable size, the Forth produces no less than 1,538, considerably more than half its total or 19 per haul, St Andrews Bay giving 859 or 28 per haul. The maximum occurred in August in the Forth, and in St Andrews Bay in March; the minimum in March and December (76) in the Forth, and in December in St Andrews Bay (11). Of plaice from 7—11 inches the enormous numbers in St Andrews Bay are clearly brought out, for whereas 79 hauls in the Forth captured but 854 or 10 per haul, 30 hauls in St Andrews Bay gave 3,472 or 115 per haul—eleven times as many. Of this great number 2,464 pertain to August, with only five hauls of the trawl, whereas six hauls in the Forth gave but 162 in the same month, and five hauls in September only 10. Plaice under 7 inches were rare in the Forth—15 occurring in 79 hauls, and chiefly in the first half of the year. On the other hand, 737 were procured in the 30 hauls in St Andrews Bay, and all but 12 in the month of August. The broad features distinguishing the two areas are nowhere mapped out more decisively than in the case of this fish, and if the Fishery Board had been able to devote as much energy to St Andrews Bay as to the Forth the results would have been even more pronounced. No feature was more evident during the work off the east coast of Scotland in 1884 than that the large plaice were as a rule offshore fishes, whereas the sandy shallow water inshore was a stronghold for the smaller forms.

The total number of dabs of all kinds was 2,276, or 28 per haul, an increase on the previous year, though one under 1886, but little real importance is to be placed on the variations exhibited under the circumstances by this species. Most were captured during the latter half of the year, that is, from July to December.

This year cod showed a larger average than had yet been obtained, viz. 23 per haul, or a total of 1821 of all kinds. Unfortunately, the statistics have not been uniformly kept throughout the various years, the separation of the large cod (23—27 in.) having only been clearly carried out in 1890. So far as can be observed, however, a tendency to increase in the colder months in the case of the larger cod is evident. The highest monthly number of all kinds is in November: December follows closely, then October. The highest monthly number of large cod (23—27 in.) is in February,—November, October and December following. During the five years, the averages of all kinds of cod were 8, 12, 10, 10, 23; the sudden leap in the last year (1890) being one of the variations so prevalent in the department.

Long-rough dabs had a total of 1,627 or 20 per haul, a

larger number than the two previous years, the averages from 1886 being per haul—15, 28 (1887), 13 (1888), 16 (1889) and 20 (1890). About one-half were saleable in 1890. The species forms a favourite food of many fishes, and its apparent increase is not objectionable.

The important lemon-dab shows a tendency to diminish during the five years. Thus in 1886 the average per haul was 16; in 1887, 47; in 1888, 17; in 1889, 11; and this year only 9 per haul, or a total of 771. It is not an abundant form, though generally distributed, and frequents ground, as a rule, unfavourable for trawling. It is often caught by the liners, and in St Andrews 30 years ago was less esteemed than a plaice of the same size. No great weight is to be attached to the apparent diminution.

The grey gurnard had the same average, viz. 9 per haul, the lowest that had occurred in the series, but such small numbers may be due as much to the habit of the fishes as to any marked general diminution in their numbers.

All the stations showed an increase on the previous year except Station IV. (the line along the shore towards Gullane Ness). The average increase per haul (saleable and unsaleable) of the food-fishes found on each from I. to IX. is as follows:—increase of 36, 78, 142, diminution of 133 (IV.), increase of 128, 3, 179, 197, 78.

That with an increase of 27 hauls in 1891 the total number of fishes should be reduced to 18,012 or less by 258 than in 1890 is sufficient proof that the work was less productive. It is true that an additional winter month (January) handicaps 1891, but this is more or less balanced by the captures in June, a month absent in 1890. The average per haul on the total is 169 or 62 less than in 1890. 13,325 or 125 per haul, were saleable, and 4,687 or 44 per haul, unsaleable. Every month of the year was represented, though not all by an equivalent number of hauls.

The previous year the haddock gave way to the whiting, but this year both yielded to the plaice, so that it might be argued that here we have an instance of the nomad round fishes being diminished by over-fishing in the open waters, while the less pelagic flat fishes have increased in the protected area and its vicinity. It is unsafe, however, to theorise under the circumstances. The total number of plaice was 3,723 or 35 per haul, and only 18 were unsaleable. In grouping the plaice as in the previous year according to the months, similar features

1891.

Plaice.

No. of Hauls		19-26 in.	12-18 in.	7-11 in.	Under 6 in.	Total	Average per Haul
9	Jan.	2	101	17	—	120	13
9	Feb.	—	142	76	3	221	24
7	Mar.	3	243	54	—	300	42
4	Apl.	—	22	7	1	30	7
14	May	4	264	375	10	653	46
9	June	4	234	331	2	571	63
9	July	2	188	276	1	467	51
9	Aug.	2	177	124	1	304	33
7	Sept.	2	118	168	—	288	41
11	Oct.	9	259	57	—	325	29
9	Nov.	15	207	101	—	323	35
9	Dec.	—	112	6	—	118	13
106		43	2067	1582	18		

are brought out as to sizes. Thus of large plaice one occurs in every 2 or 3 hauls. Of plaice from 12—18 inches, the common size in the market, precisely the same proportion was present, viz. 19 per haul—a noteworthy feature. The maximum this year was in June, whereas last year it was in August, the minimum in April. Of the next size 7—11 inches, an increase appears, viz., 14 per haul instead of 10 in 1890. The maximum was in the same month (June), whereas both were in August in 1890, a fact which shows how futile it is to attempt to draw deductions of importance from these details. It may be that the weather enabled the trawl to work with special success on these occasions, or that a particular series of commotions or currents affected the habit or food of the species.

Those under 7 inches were even rarer than last year (18 in 106 hauls). So many factors come into play in making comparisons between years that it is extremely difficult to attain reliable results, though occasionally an interesting deduction is made, *e.g.* that of the medium plaice (12—18 inches) in the Forth in 1890 and 1891, yet it may be only a coincidence.

The next in number are the dabs with a total of 3,547 or 33[1] per haul. Of these 1,865 or 17 per haul were saleable and 1,682 or 16 per haul unsaleable. A similar increment to that of the plaice has therefore occurred on the figures of last year, though no special weight need be put on this variation. The

1890.

Dab.

No. of Hauls		12–15 in.	7–11 in.	Under 7 in.	Total	Average per Haul
	Jan.	—	—	—	—	—
9	Feb.	—	18	3	21	2
9	Mar.	4	55	7	66	7
9	Apr.	—	38	20	58	6
9	May	1	118	48	167	18
	June	—	—	—	—	—
7	July	3	236	147	386	55
6	Aug.	—	94	345	439	73
5	Sept.	1	120	92	213	42
9	Oct.	1	298	191	490	54
7	Nov.	—	129	72	201	28
9	Dec.	—	121	114	235	26
79		10	1227	1039		

1891.

No. of Hauls		12–15 in.	7–11 in.	Under 7 in.	Total	Average per Haul
9	Jan.	1	26	21	48	5
9	Feb.	—	101	84	185	20
7	Mar.	—	100	25	125	17
4	Apr.	2	38	30	70	17
14	May	1	238	116	355	25
9	June	4	229	233	466	51
9	July	4	153	212	369	41
9	Aug.	2	165	173	340	37
7	Sept.	1	150	75	226	32
11	Oct.	4	324	393	721	65
9	Nov.	16	138	226	380	42
9	Dec.	1	167	94	262	29
106		36	1829	1682		

increment, moreover, was general, and not by a sudden rise of a special size or stage throughout the season, indeed, the highest average per haul (73) occurs in August, 1890, the next being

[1] Fractions as usual are omitted.

65 in October, 1891. Of the three sizes the following is the result. A large dab appeared about once in eight hauls in 1890 and about 1 in 3 in 1891; the average for the medium in 1890 was 15 per haul; in 1891, 18 per haul; the average for the unsaleable (under 7 inches) for 1890 was 13; for 1891, 15 per haul. The high average of August 1890 (73) was composed mainly of those under 7 inches, whereas the highest in 1891 (65) was more equally divided between medium and small (324—393). No important deduction can be made from such variations. They are associated with all fishing operations.

The total for whitings was 2,302 or 21 per haul—less than a third what it was in 1890. Of this number 1,976 or 18 per haul were saleable and 326 or 3 per haul were unsaleable. By some this would be held as an evidence of the decadence of the whiting-fishery, yet it is 6 per haul more than were captured in 1887, the year of plenty in the "Garland's" returns, and more than twice what it was in 1886. But the curves of this species from year to year show how unsafe it is to draw conclusions from such changes. The maximum occurred in October, which had a total of 617, not half the total (1,525) of the same maximum month last year.

Long-rough dabs followed with a total of 2,242 or 21 per haul, the next highest to the year 1887, which had an average of 28 per haul. They agreed with the dabs in having their maximum in October. The saleable numbered 873 and the unsaleable 1,369.

The gurnard is a fish which is generally caught in moderate numbers by the trawl. This year the total number was 1,638 or 15 per haul, exactly the same as in 1887 when all the figures were so high. Persistent trawling outside and line-fishing both there and in the closed waters would probably have a tendency to reduce such a species as well as the haddock, yet here we have a form, for which perhaps fishermen care but little, keeping up its numbers with a steady persistency, notwithstanding the means for its destruction are the same as those for the haddock. On the other hand, it may be said that the gurnard as a rule swims higher than the haddock, and is less apt to come in the way of the trawl. The fact that they are caught in considerable numbers by the trawl would show however that they share the same danger, and are to be judged by a similar standard.

The total for the haddocks was 1,435 or 13 per haul, the lowest on the list for the ten years. Of this number 1,301 were saleable and 134 unsaleable. The contrast between this year and 1887 is marked, since the amount is only about $\frac{1}{10}$th of what it was in the latter. It would be unsafe, however, to say that the haddock was disappearing, since within a year or two the waters of the east coast swarmed with young haddocks, to the annoyance of the fishermen, and the "Garland's" captures four years subsequently had an average of 2 per haul above those of 1886.

The total for cod was 1,229 or 11 per haul. Of this number 1,123 or 10 per haul were saleable and 106 or 1 per haul unsaleable. This was but half the number procured in 1890. While in the latter year the maximum occurred in November, this year it was in January, a feature probably due to the prevalence of herrings or other food in the Forth. The average was three above what it was in 1887.

Lemon-dabs were somewhat more numerous than in 1890 by about 1 per haul, the total being 1,135 or 10 per haul. Only 49 were unsaleable. Their maximum occurred in July.

In regard to the total quantities of all kinds of food-fishes on the various stations, a slight increase (3) on the previous year in the average per haul occurs at I., a decrease of 37 at II., a decrease of 108 at III., an increase of 105 at IV., a decrease of 110 at V., a decrease of 20 at VI., of 113 at VII., of 201 at VIII., and of 94 at IX.

Just as in St Andrews Bay, the average of this year was low, the second lowest indeed in the whole series in the Forth. The causes of this depression to some extent lay in the inclusion of all the winter months (every month of the year being represented), and probably also to the weather, mode of working and state of the tide, as in St Andrews Bay.

During the year 1892 a new station (X.) was added to those already existing at the mouth of the Forth, but no mention is made of it in the Report of the year (Eleventh). It has been deemed unnecessary to refer to this station, since it would only confuse and complicate the statistics dealt with from the beginning and afford no further information.

Ninety-nine hauls of the trawl gave a total of 16,736 fishes of all kinds or 167 per haul, a result very near (two less) that of last year, yet the month of August was omitted, so that to balance the six winter months of January, February, March, April, November, and December there were only five productive ones, and of these October had an unusually small total, viz. 673, as against, for instance, 1,571 last year. Of the total number for the year, 9,394 or 93 per haul were saleable, and 7,342 or 74 per haul unsaleable, a different arrangement from that of 1891, in which the saleable forms were nearly three times that of the unsaleable.

The plaice and the dab, as last year, hold the foremost positions in regard to numbers, though the haddock, by exactly doubling its average per haul on last year, follows closely, and once again illustrates the uncertainty in all fishing operations.

The total number of plaice in 1892 was 2,806 or 28 per haul, a reduction of 7 per haul on the previous year. Of this number only 44 were unsaleable, a satisfactory condition of things in inshore water. The large plaice, as usual, were few in number (32), the medium most numerous (1,485), and the saleable small (7—11½ in.) 1,245. If, in ordinary trawling, the latter, as in the case of the "Garland," were returned to the water their value would soon have greatly increased.

Of the 2,717 dabs (27 per haul), 1,095 or 11 per haul were saleable, and 1,622 or 16 per haul unsaleable.

Of haddocks there were 2,637 or 26 per haul, 1,115 or 11 per haul being saleable, and 1,522 or 15 per haul unsaleable. The latter were chiefly procured from September to December, the maximum being in October, and probably represent the young haddocks of the season making for inshore waters from their earlier haunts in the deep water. The uncertainties connected with fishing operations are shown in the disproportions between the saleable and unsaleable of this year and the preceding, where only 134 out of 1,435 were unsaleable.

Like the dabs, the majority of the long-rough dabs were unsaleable. Thus out of a total of 2,274 or 22 per haul, 782 or 7 per haul were saleable and 1,492 or about 15 per haul unsaleable. The two species just mentioned have always held a prominent place amongst the unsaleable forms in trawling, but neither is a fish of great importance in the market, though dabs are excellent food. They take a conspicuous position, however, as the food of other fishes. In most respects the captures of the long-rough dabs coincided with those of the previous year.

Gurnards reached the highest number yet met with in the Forth, viz. a total of 1,944 or 19 per haul. Of this number 1,234 or 12 per haul were saleable and 710 or 7 per haul unsaleable. There was no sign of any special accumulation of such a species by the closure, for it varied much both before restriction and subsequently[1].

The total for the whiting was 1,898 or 19 per haul—two below last year. The saleable were 740 or 7 per haul, and the unsaleable 1,158 or about 11 per haul, in both cases contrasting with the condition the previous year, in which only 3 per haul were unsaleable and 18 per haul saleable. The maximum of the saleable whitings was in January, but, as for the last two years, the maximum for all sizes was in October, this being due to the capture of 505 unsaleable young forms of the season as they were seeking the inshore waters.

There were 951 lemon-dabs or 9 per haul, only a fifth of the number captured in 1887, and one under last year. They were nearly equally divided between saleable and unsaleable. The total for cod was 841 or 8 per haul, the same as in 1886, and the lowest figure on the list. All were saleable, and 105 were large cod. A single sea-trout of 9½ in. was captured. Single examples of this species, as a rule, when about 10 or 11 inches in length, are procured in herring-nets—sometimes 25 miles or more from land.

Most of the stations showed a reduction in the average per haul. The following had an increase, viz. V. of 61, VI. of 7, VII. of 11, VIII. of 32, and IX. of 21.

[1] One of the somewhat numerous slips in the S. F. B. Reports has it that "Gurnards" showed a considerable falling off (1892, p. 25).

With the exception of 1890 the series of years following the fertile one of 1887 had given but indifferent totals for the food-fishes of the Forth. Four of these had been under that reached at the commencement of the experiments, while 1890 had an average of only twelve over it. The sudden rise of 160 per haul in 1887, and which had buoyed so many hopes, therefore, had been ephemeral—dependent alike on the short and favourable period of the year and on other advantages. In the year 1893 the grand total was three times larger than in 1887, viz. 27,401, but the hauls which produced it were four times as numerous, viz. 96. The average per haul was 285, or 93 under 1887, yet it came next it during the period of eight years. Of this number 14,130 or 147 per haul were saleable and 13,271 or 138 per haul unsaleable. The average of the saleable had been exceeded in 1886, 1888, and 1890, whereas that of the unsaleable was exceeded only in 1887. The years 1887 and 1893 were therefore remarkable in the capture of unsaleable fishes, the former having the largest proportion in the series, the latter also a large proportion, but greatly under that of 1887.

All the months of the year are represented in the table, though unequally, March being lowest with only two hauls, July and November with 7, January, February, May, and September with 8, April, June, October, and December with 9, while August had 11. By far the most productive months were August and September. The work was thus very fairly distributed over the year, but the diminution of the hauls in March would tend to keep up the general average since it is usually an unproductive month, while the 11 hauls in August would also tend to raise the average.

No fish has been more prominently brought forward by the liners and others as an instance of the gradual decay of the Scotch fisheries than the haddock. Statistics, in some cases most difficult to comprehend, have been produced to bear out such views, and they have been accompanied by the most lugubrious pictures of the present and future of the line-fishermen of our shores. But such advocates, whose real interest in the fishermen—it is to be hoped—is to be measured by the warmth of their efforts, hitherto have not been able to satisfy those who have gone minutely into the subject, and hence have been met with doubt and a desire for further information. The main object of the various Royal Commissions, of the Scotch Fishery Board with its extensive official ramifications and its large expenditure, of the special investigations which have, within the last 14 or 15 years, been made on the life-history and development of the food-fishes of our shores, is the welfare of the fishing industry in all its bearings. To promulgate laws for the administration and regulation of this department—without an adequate and practical acquaintance with the actual facts of the case—would be most unsatisfactory if not disastrous, and unworthy of a country in which the search for truth and the advancement of knowledge are sufficient at all hazards to enlist its best energies.

The foregoing remarks have been suggested by the position of the haddock in the experiments of 1893. Beginning in 1886 with the fair average of 84, then rising to the high average of 129 in 1887, it for five years subsequently had persistently kept at a low average, indeed, in 1891 at a very low average (13). There was thus a basis on which to rest the melancholy pictures which for many years had been drawn of the Scotch line-fishing. Here, it might be said, was clear proof that the haddock was diminishing in the Scottish seas, and that protection of one kind or other was clamant. But nature, in the vast domain of the ocean, rises above all the theories of the scientific and above all the notions of the "practical" man. This year (1893) the grand total of the haddock, which we were to believe had been swept out of every bay and fishing-bank on the eastern shores of Scotland, mounted in itself above the total of every kind of fish caught in the experiments of 1886, 1887 and 1888, and was only 268 less than that of 1889. This single year produced in the experiments as many haddocks as any five of the preceding. Fishermen were, indeed, struck by the unusually large quantity of small haddocks both in the inshore and offshore grounds along the whole of the east coast, and which made their appearance in July (1893). Of this total of 12,500, or 130 per haul, the highest average yet attained for this fish in these experiments, extending over 8 years, 3,978 or 41 per haul were saleable, and 8,522 or 88[1] per haul unsaleable.

It might be said, however, that if we glance at a table (viz. Table XI.) of captures of haddocks from 1886 onward, that at any rate the saleable haddocks had greatly diminished—since the average in 1886 was 77 per haul, while subsequently it had never reached more than 44 (1888), and had been as low as 11 (1892). But the same argument is applicable to the unsaleable, in which the range is even greater, viz. from 1 to 112, and, besides, the accurate separation of the saleable from the unsaleable was not carried out in 1886 on the same lines as subsequently. The captures show no lack of haddocks, but demonstrate the proverbial chance of the pursuit.

In another way the contrast with 1887, for instance, may be brought out, viz. by calculating the average of the haddocks caught in the colder and in the warmer months. Thus in the former 1899 or 44 per haul were procured, while in the latter no less than 10,581 or 199 per haul were obtained. When weighed in the same balance, therefore, the prominent position assigned to 1887 disappears.

The average for last year had been but a fifth of the number, and yet, without preliminary warning, this great increase demonstrated the insecurity of the position held by the pessimist, and showed that now, as in all previous times, there are few departments so full of surprises as in the fisheries. A consideration of the history of the herring-fishery, for instance, is an excellent illustration, and it is still more striking in those fishes with pelagic eggs. July, August and September were the months in which the haddocks were in greatest numbers—both saleable and unsaleable. In regard to size, comparatively few, only 22, were large, 231 being medium, and the small saleable (from 8 to 9 inches) formed the majority (3,727). The large haddocks were obtained chiefly in the colder months, probably in connection with the advent of herrings in January and February.

The next in numbers is the dab—3,623, or 37 per haul, the second highest in the 8 years. Of this number 1,440 or 15

[1] It has to be borne in mind that decimals are avoided throughout.

per haul were saleable and 2,183 or 22 per haul unsaleable. Plaice follow with a total of 3,362 or 35 per haul, all being saleable except 46. Comparatively few (44) large plaice (over 19 inches) were, as usual, obtained, though the proportion is larger than in St Andrews Bay. Long-rough dabs had a total of 2,105 or 21 per haul, of which 935 or 9 per haul were saleable and 1,170 or 12 per haul were unsaleable. This fish seems to present in these tables a more equable average than most forms throughout the period of eight years; moreover, its monthly averages do not as a rule show the accentuation of the warmer months so prominently as some species.

The total number of gurnards was 1,497 or 15 per haul, of which 1,093 or 11 per haul were saleable and 404 or 4 per haul unsaleable. Thrice during the period the average had been the same. The maximum was in May, and the warmer months had higher numbers. The total of the lemon-dab was 1,255 or 13 per haul, of which 1,119 or 12 per haul were saleable and 136 or 1 per haul unsaleable. Their numbers were highest in the warmer months, though they occurred in considerable numbers every month. Of cod there were 1,113 or 11 per haul—all saleable. Whitings followed with 1,030 or 10 per haul, a comparatively small number, though one more than in 1886.

Without exception every station showed a decided increase of the average on the previous year, the greatest being at Station VII. west of the Bass, where it was no less than 528 (saleable and unsaleable) per haul, a total of 314 over that of the previous year, an increment which, in itself, exceeds all the previous hauls on the station except in 1887, when it was only 44 higher (358). Station V. had an increase in the average per haul of 186 over 1892, and exceeded that of 1887 by 10. Station VIII. had 142 over the previous year, and exceeded the average of 1887 by 24. In the others the increase was decided, though smaller.

The year, therefore, may be regarded as a very successful one, and, though it may not lend much support to the idea of accumulation, it does not countenance the notion of gradual decadence of the fisheries. The month of December, in which 9 hauls took place, showed a somewhat high average, viz. 117 per haul, as compared with 114 in November. Such was chiefly due to the increased number of small cod and haddocks.

In 1894 seventy-one hauls of the trawl were made in the Forth during nine months of the year, six of these, however, belonging to the colder series, viz. January, February, March, April, November and December, and only three pertaining to the warmer, viz., June, August and September. The apportionment of the working season therefore must have influenced the results, since the productive months of July and October were wholly absent. These 71 hauls captured a total of 19,581 fishes, the third highest in the series, the average per haul being 275—which held the same position. A striking feature, as in 1890, was the large proportion of saleable fishes, viz. 15,349 or 216 per haul, a number exceeding any record during the period, being 27 over 1890, 81 over the successful year 1887, and 69 over the preceding year (1893). The unsaleable were 4,232 in number or 59 per haul, a very moderate number, and no less than 184 per haul under the unsaleable average of 1887.

The influence of the colder months in affecting the average of a year is clearly observed by taking the four warmer months, viz. April, June, August and September, and contrasting them with the average for the five colder, viz. January, February, March, November and December. Although April is usually classed with the colder months, it may be included on the other side in the present instance, so as to give a fair field. The number of hauls in the warmer months was 36 and the average was 370 per haul[1], a similar figure to that (378) obtained in 1887, when the favoured work was carried on in July, August and September. If the five winter months of 1894 are now considered, the average of the 35 hauls is not half the foregoing, or only 177 per haul. Nothing more conclusively shows that, in dealing with such questions, the utmost

[1] The 4 warmer months had each 9 hauls, the 5 winter were less regular, February and March had each 9, January and November each 6, and December 5.

caution is necessary. It is easy to raise hopes by new measures in a struggling industry: it is less easy to allay the disappointment entailed by their failure.

The haddock again outstrips all the other fishes in the returns for 1894, the total being 6,681, or an average of 94 per haul, the third highest in the ten years' experiments. Of this total, 5,841 or 82 per haul were saleable and 837 or 11 per haul unsaleable. Not only, therefore, was the general average high, but the average of the saleable was the highest in the ten years. If we compare the haddocks captured in the warmer months, as given above, with the colder, the average per haul in the former is 130, or a higher average by 1 than obtained in the very productive months of June, August and September in 1887. The five colder months give an average of only 56.

The next in regard to numbers is the dab, the total for which is 3,111 or 43 per haul, a comparatively high average, six over the preceding year and only five behind that of 1887 to which it comes next. Of these 1,903 or 26 per haul were saleable and 1,208 or 17 per haul were unsaleable. Such a rise is one of the vicissitudes constantly met with in the fisheries, and little weight is to be attached to it further than that it indicates the resources of nature.

Plaice follow with a total of 2,624 or 36 per haul, an average of only one more than last year, and three below what it was in 1886. Of this number all were saleable except 53—in this respect resembling the haddock of the same season.

Long-rough dabs had a total of 2,087 or 29 per haul, the highest average attained in the period (9 years). Of these only 641 or 9 per haul were saleable, whilst 1,446 or 20 per haul were unsaleable. This is not a fish of much importance in the market, though palatable enough, and it is of small size as a rule. Other fishes prey on it to a large extent. Whitings had a total of 1,746 or 24 per haul, more than double the average of the previous year. Of these 1,660 or 23 per haul were saleable, the majority being between 7 and 9 inches, and only 14 over 14 inches, 574 being between 10 and 13 inches. Only 86, or 1 per haul, were unsaleable. The numbers of the cod exceeded those of 1893, the total being 1,152 or 16 per haul, considerably over that of 1887, though less than in 1890. Of this number 175 were unsaleable, that is, under 8 inches in length. The total for lemon-dabs was 891, or 12 per haul, a moderate number; 97 were unsaleable. Gurnards were comparatively few, viz. 9 per haul, and a similar number (10 per haul) was obtained in St Andrews Bay, where usually they are more numerous than in the Forth.

Of the nine stations five showed a diminution, viz. I. of 7 per haul, VI. of 41, VII. of 179, VIII. of 61, and IX. of 45; while four showed an increase, viz. II. of 11, III. of 14, IV. of 52, and V. of no less than 202. It is noteworthy, however, that no hauls were made on V. in January, nor in December, and thus it was relieved of two of the non-productive months. High figures for haddocks during the other months thus brought the average of this station, which lies in the central channel of the Forth, almost to the top of the list, for the high figure of 613 per haul has only once been exceeded in the series, viz. by 616 in 1887, at Station II. (from Dysart to Wemyss). The comparison, however, is unreliable, since the work in 1887 was carried on only in June, August and September. Small haddocks, however, in this instance also swelled the totals.

The feature of the year was the very large proportion of saleable fishes, by far the highest yet met with in the series. Moreover, the fact is brought home that by omitting to work on a station (*e.g.* from storms or otherwise) once or twice during a season the comparative values are materially altered, and are apt to mislead rather than assist in determining data. It would have been well to have rigidly adhered as far as possible to the regulation number of hauls on each station during a particular month.

As formerly mentioned, little weight is placed on the comparison of the closed with the open area, either in regard to round or flat fishes or totals, especially as no note is taken of the particular months when work was carried on in each year. The working of the trawl in 18 to 20 fathoms on board the "Garland" would be different from that in shallower water, and the fact that in 1886 the two areas gave such different results shows that the circumstances of the two differ. The number of hauls was also much fewer in the open area. Lastly the irregularities of the totals, caused it is true by the sudden increase of round fishes, show the usual uncertainty of the pursuit. Thus, in 1893, the average of all fishes rose, according to the Blue-book, in the open area to 267, a figure only twice exceeded in the closed area. Further, while in the closed area, the flat fishes, which have their haunts on the sandy and other grounds in the estuary, always exceeded those from the open water, the round fishes reached in 1893 in the open area the highest average of the series, the next being in the same area in 1890. It would, indeed, have been a satisfactory solution of the difficulty to have found the food-fishes of the Scottish shores substantially and surely increased by the closure, but a perusal of the tables on p. 19 of the 13th Annual Report of the Fishery Board, and more especially the remarks in the concluding paragraph on the Frith of Forth after nine years' work is disappointing:—In the closed area "the tables show that among flat fish there was an increase of all kinds, except lemon soles, turbot and brill. The greatest increase was in long rough dabs."..."Among round fish there was a decrease in haddocks and gurnards, and an increase in whitings and cod." "In the open area, the decrease in the flat fishes was common to all kinds except dabs, and there was a decrease in haddocks and gurnards." An increase due to the abundance of dabs and long-rough dabs is not of much importance in either area; but what is of more moment is the demonstration of the fact that these returns from first to last in both open and closed areas maintain on the whole a fairly steady character. It is difficult to see, however, what relation the closure bears to the averages.

In 1895, the last year of the experiments in the Forth, 75 hauls of the trawl were made in the Forth during ten months, five being colder months, viz. January, February, March, April and December, and five warmer months, viz. May, July, August, September and October. The total number of fishes captured was 20,570 or 274 per haul, a comparatively high average, sufficient to meet the statement of diminution, and yet not sufficiently high to support the theory of gradual accumulation. Of this number 16,960 or 226 per haul were saleable, the largest entry during the ten years, and 3,610 or 48 per haul unsaleable, one of the lower entries in the list and no less than 195 under the unsaleable of 1887. These figures would appear to give no support to the view that excessive trawling in the offshore grounds has steadily diminished the food-fishes of the Forth. The year was a fair test in regard to periods and equality of hauls, and the result is therefore of special interest.

During the five colder months 33 hauls of the trawl occurred, producing a total of 3880 fishes or 117 per haul, whilst the 42 hauls in the warmer months gave 16,690 or 397 per haul, a sufficient corroboration of the divergent conditions of these periods. Such also demonstrates how misleading it is to use the statistics of 1897, which had no colder month at all, on the one hand to illustrate the immediate influence of the closure, and on the other the gradual deterioration of the inshore waters since that date by the unrestricted fishing in the waters beyond.

The preceding facts are brought out quite as clearly in the case of the haddock, which, as usual, is at the head of the list of captures of the year with 7,033 or 93 per haul, only one under last year. This average appears to be much below that of 1887, in which haddocks were 129 per haul, and, therefore, ostensibly forms a basis for demanding further closures so as to control the "spawning grounds" of the parent-fishes, which are being devastated by the trawlers, and general free fishing in the open waters. But if we contrast the numbers caught during the five colder months, viz. 435, with those of the warmer months (which alone formed the working period in 1887), viz. 6,598, or 13 per haul against 157 per haul, some hesitation may reasonably be felt in recommending such methods of controlling the important subject of the fisheries of the country. Not only is the average for 1895 found to be higher than in 1887, but that of 1893, for instance, exceeds them both, for its average in the warmer months was 199. In glancing over the 10 years' work in the Forth, there is no ground for any other view than that of confidence in the supply of the haddock. Years occur in which the captures are very small, as in 1889 and in 1891, but they either gradually or suddenly rise to their former standard or exceed it, as has happened in every department of the fisheries of the open sea probably from time immemorial. The maximum numbers, irrespective of hauls, are found in August, which is boldly marked by 11,190 in the decade, followed by 6,959 in September, and about equally supported by October and July, the former with 6,194 and the latter with 6,175.

	Warmer months May—Oct.	Colder months Jan.—Apr., Nov. & Dec.	Total	Difference
1st period				
No. of Hauls	159	110	269	50+ in favour of warm mths.
2nd period				
No. of Hauls	214	219	433	+5 in favour of colder mths.

The haddock in 1883–84 and in 1895–96 has been specially singled out for a demonstration of the progressive ruin of the Scottish fisheries. The period of 1883 and 1884 is contrasted with 1895 and 1896, and it is stated that in the latter period there is a decrease of £42,537. 10*s*. No heed is given to the view that either period might have been exceptional, as has so often happened, for instance, in the herring- as well as the haddock-fishing. The figures, however, may be examined in passing. A preliminary statement is made that in 1883–84 "there were only a few trawlers on the coast, whose statistics were not kept separate," but this is scarcely accurate, for the powerful new ships of the General Steam Fishing Company of Granton, others from the same port and from Leith, those from Dundee and Aberdeen, besides the English vessels working in Scottish waters, were actively engaged in that period, not to speak of sailing trawlers. Indeed in the Trawling Report of 1884, the author estimated the captures by trawlers in Scottish waters as at least 9,000 tons or 180,000 cwts., an amount which would effectually extinguish most of the imaginary loss to the liners in this comparison. The average value of haddocks captured in 1883-84 is given as £320,702, but it is well also to take note of their weight:

	cwts.
1883	543,568 = £340,693[1]
1884	464,049 – £300,712
Total	1007,617

or an average for the period of 503,808 cwts. = £320,702. 10*s.* Doubtless the critic thought he had to do with a period when the distribution of the haddock in our seas was much more abundant, and, as the captures by liners and trawlers were, he says, not separated, there was a certain haziness not inconvenient for argument. But during that period accurate investigations had been made on most of the important grounds, and thus the situation was well known.

During the second of the periods the following is the condition:—

		cwts.
1895	Haddocks landed by liners	663,548 = £272,692
	" " trawlers	337,792 = 136,914
	Total	1,001,340 = £409,606
1896	Haddocks landed by liners	670,685 = £283,638
	" " trawlers	319,513 = 136,847
	Total	990,198 = £420,485

or an average of 995,769 cwts. = £415,045. That is to say, during the second period, nearly double the weight of haddocks was landed. It is thus, perhaps, well that weights were left by the critic out of sight, else the $\frac{1}{4}$ for the liners and $\frac{1}{2}$ for the trawlers (supposed to be taken outside 20 miles) would not have been of much service. Fortunately, moreover, prices do not regulate the distribution of the marine food-fishes.

The value of the haddocks of the period 1883-84, is correctly given, viz. an average for each of the two years of £320,702. 10*s.*, but it is assumed that the fishes were all caught within 20 miles, a conclusion open to doubt.

In regard to the second period, we have other sources of

[1] A writer holds that the market-value of fishes has been increased by 200 per cent. during the decade 1886—96. He will not find these statistics bear out his view. The application of his opinions to the promulgation of the theory of the dearth in the inshore waters misses the mark.

information as to the condition of the haddock in the Scotch seas than those apparently in the critic's hands. Thus, we have (1) the accurate data of 1884 with which to contrast the second period; (2) the indifferent, but, at any rate, unbiassed information derived from the work of the "Garland" over the greater part of the interval as well as the second period; (3) other observations during the period; and (4) the condition of certain areas at the present time (1898). These accurate data show that the condition of the Moray Frith, as regards haddocks in 1884, was barely equal to what the unenclosed area beyond it was in 1898 almost on the same dates (*vide* p. 207); that a single vessel in January, 1898, captured 60 miles E.N.E. from Aberdeen (and this brings it within a reasonable distance of land and of the Moray Frith), no less than 370 boxes of haddocks, besides other fishes in three hauls of the trawl; that very great variations in the captures of the haddock by the "Garland" in the Moray Frith occurred, the high and low numbers having little relation to the actual abundance or scarcity, as proved, for instance, by the captures of the liners in the area. The same fish in St Andrews Bay is much as it was in 1884, now with a high average, now with a low average, and the uncertainty of this fishing is shown by the sudden variations[1]. The reduction of the average in the second five years is not due to any change in distribution, but to other causes elsewhere explained[2]. In the same way the condition of the haddock in the Forth remains in 1895 as it was in 1886, for little weight need be placed on the increase (25 per haul) in the latter five years of the decade. The circumstances under which the work was carried on in the two quinquennial periods were different, as explained in its proper place (p. 183).

[1] Not only do the captures of haddocks vary, but the opinions of commercial men ("practical men") likewise vary. It is not so long ago since one of them forwarded from the deeps off the east coast a series of stomachs of large haddocks distended with the spawn of the herring, as a proof of the destructive tendencies of these fishes, and of the great benefit to the herring-fishery the capture in the trawl of about 80 boxes of these must have been.

[2] *Vide* p. 125.

Haddocks for a time may be scarce, but generally they appear in similar numbers or in greater abundance than before, and there is nothing in the whole history of this fish for hundreds of years that would lead to the conclusion that any great change by way of reduction has occurred.

To meet the foregoing it is necessary for the critic to resort to the arbitrary and very uncertain method of subtracting no less a sum than £137,981 from the average amount of the period 1895 and 1896, for haddocks caught beyond 20 miles from shore. But he ought first to have proved that there were no haddocks to be captured within the 20-mile limit. It will not do to reply that large captures were made at the greater distances. That is sufficiently known, else the ships would not repeat the experiment. If these ships had chosen to work within the 20-mile limit they would have captured by skill and energy a large number of haddocks which have been proved to be there, and their fresh condition on landing would have enhanced their value, so that this method of argument is dubious. Besides, the records of haddocks caught beyond 20 miles from shore, on which the calculation is made, have not been produced. If the liners do not work the enclosed waters for haddocks with the same pertinacity as the trawlers outside, that is not to be made the fault of the waters. Certain fishes are there to be captured by every legitimate method of fishing, and there are few—with the real interests of the fishermen and of the country at heart—who would, after careful deliberation, restrict the method to line-fishing, which remains very much to-day what it was centuries ago. If some enclosed waters are only occasionally fished, and that with little energy and enterprise, the question as to whether the country gains all the benefit from such areas that it should at once suggests itself.

Plaice come next with a total of 3,299 or 44 per haul, an increase of 8 per haul on last year. Of this number almost all were saleable, only 33 not being of this category. Though the number caught during the warmer months is much higher than during the colder months, the proportion per haul is different from that in the haddock. Thus, 1,115 were caught in 33 hauls during the colder months or 33 per haul, while 2,219 were taken in the warmer months or 52 per haul. Contrasted with 1887 in which a working period of only three warm months occurs, it is found that the average of plaice per haul in that year was 73, the difference in favour of the latter being 21. Here, then, is a case of diminution which the advocates for restriction might legitimately have brought forward. On going over the ten years and separating the hauls during the colder months of January, February, March, April, November and December, and during the rest of the year (warmer months), as in the accompanying table, it is found that the columns for the

FORTH.

Plaice, average per Haul.

	Colder months	Warmer months
1886	19	44
1887	—	73
1888	2	58
1889	33	51*
1890	18	42
1891	23	44
1892	20	35
1893	28	40
1894	31	45
1895	33	52

* Same no. of months and same no. of hauls in each (45).

colder and the warmer months show considerable variation, but it is clear that they stand at the end of the period better than they did at the beginning by 14 on the colder months and 8 on the warmer months. 1887 was a proportionally high year, but there is no reason to think that the resources of nature have been in any way materially altered throughout the period. The usual variations took place, the catch, for instance, in 1887 greatly exceeding that of 1892, but the supply steadily kept up, both in the warmer and the colder period from first to last.

As in the haddock, the maximum monthly number of plaice occurred in the decade in August, viz. 4,465, the next highest, 3,426, being in September; July follows with 2,844, and June with 2,400, then May with 1,982, whilst January and December are lowest.

Dabs follow with a total of 2,748 or 36 per haul, seven under that of last year. Of this number 1,751 or 23 per haul were saleable, and 997 or 13 per haul were unsaleable. This form has on the whole kept up fairly throughout, now rising now falling in the usual manner. It cannot be said that unrestricted fishing outside steadily affected this species, nor is there evidence of accumulation. Beginning with an average of 29 per haul in 1886 it ran up suddenly to 48 in 1887, then next year was reduced to exactly half the number, and in 1889 to three less, then the average improved to 33 in 1892, fell next year to 27, rose to 37 in 1893 and to 43 in 1894. Ingenious arguments explaining the increase of the ubiquitous dabs and long-rough dabs as due to their facility in escaping where the more valuable plaice cannot pass through the meshes of the commercial trawl find no support from these investigations. The dab holds its own, irrespective of line or trawl, both in the enclosed and in the open waters of the British seas.

During the decade, the highest numbers occurred in September, viz. 4,740, though August was very little behind, viz. 4,678. It is interesting to notice the gradual rise from January to September, then the decadence in October, and the rapid fall to December. These variations are seasonal and present no feature attributable to the influence of man.

The long-rough dab comes next with a total of 2,176 or an average of 29 per haul, the same as last year. Of this number 861 or 11 per haul were saleable, and 1,315 or 17 per haul unsaleable. Beginning in 1886 with an average of 15 per haul, it suddenly rose in 1887 to 28 per haul, then sunk the following year to less than half (13), and again gradually rose with a little irregularity to 29 per haul. A consideration of the irregularities in the working-periods of the first five years and of the second five years respectively, of the fact that there was a decrease of this species in the second five years in St Andrews Bay, and of its ubiquitous distribution in inshore and offshore waters, would not lead to the conclusion that the slight increase in the second five years in the Forth was due to the small size of the species, enabling it to escape through the meshes of the trawl. The captures throughout, indeed, are characterised by the usual variations, and it would tend to insecurity to interpret them otherwise. There is a further feature in connection with the long-rough dab which is worthy of note, viz., that though the maximum monthly captures occur in September, as in the case of the dab, yet this month and August are less prominent than in the dab and plaice, and the decrease on the one hand to January and on the other to December is less pronounced.

Whitings had a total of 1,526 or an average of 20 per haul, a diminution of 4 on the previous year, though, with the usual irregularities, it was double that of 1893. Of this number all except 13 were saleable. The variations in the captures of this predaceous pelagic fish during the ten years are characteristic. In 1886 there were 9 per haul, in 1887 only 15 per haul as contrasted with 129 of the haddock, showing how difficult it is to get uniformity in regard to abundance and scarcity, even under favourable circumstances. The captures were nearly doubled next year, then fell to the same figure (15) in 1889. A sudden rise to 70 per haul characterised 1890, the most productive months being September and October. Next year the average was 21, decreasing the two following to 19 and 10, then rising to 24 in 1894, and falling to 20 in 1895. It would be difficult to draw definite conclusions from such statistics other than that the abundance of the species was little affected by the closure. The whiting is an uncertain capture at all times, it being generally supposed that it often swims higher in the water than the haddock, and thus escapes the trawl. Upwards of 1,500, besides other saleable fishes, have been captured in a single haul, and again near the same ground only a few would reward the labours of the fishermen. The appearance of the whiting in the summary of the months of the decade differs from that of the haddock since its maximum is in October (3,633). Both August and September, however, are high (2,721 and 2,298). Like the haddock the lowest numbers occur in March, while they keep proportionally high in November and December.

The total for cod was 1,147 or 15 per haul, an average of one under the previous year. All were saleable except 56. The average is 3 above the favourable year of 1887, and third in the list, the highest having occurred in 1890, when 23 per haul were captured. The small trawl of the "Garland" was little adapted for successful cod-fishing, the more so as work was carried on solely by day. In the 14th Report a slight increase (of ·7) is stated to have occurred in the last five years of the period, but this is just what would have been expected with so great a proportion of the hauls during the winter months when cod specially abound in the Forth and its neighbourhood. Consequently, high figures for this fish are found in November, December and January. A glance at the monthly periods and hauls (Table XI.) will show the essential difference between the two periods of five years in this respect. If it had been said that out of the three highest monthly totals two of these occurred in the latter five years, and thus was an evidence of increase of the species following the closure, perhaps the argument would have been more convincing. As it stands, the influence of the closure on this species in the Forth is not appreciable. Like other fishes, constant pursuit tends to render it more wary, and to break up the shoals in the nearer waters, but the ubiquitous distribution and the enormous powers of reproduction, together with the curious migration of the very young cod to the laminarious region, seem to be sufficient to enable the species to hold its own.

In the monthly totals (Tables XI. and XII.) during the decade of the "Garland's" work in the Forth, January commences with a proportionally high figure (990), higher, indeed, than any of the previous fishes (viz. haddock, whiting, plaice, dab, and long-rough dab), the numbers then decrease to May, rise to August (1047), slightly diminish to November (858), and rise in December to 1028, only 19 less than in August, a unique condition amongst the important food-fishes. The attraction offered by the herring in winter thus makes a noteworthy change in the abundance of the cod. It is also equally evident that there is a considerable diminution of this species in the Forth at its spawning-period, when it apparently seeks the waters beyond the Island of May and further outwards. The closure of the Forth does not appear to influence the abundance of this species, and it is doubtful if any extension of the limit, *e.g.* to 13 miles, could affect its distribution.

The cod is one of those forms which suffers in its young condition (codling) from the liners in the Forth. Thus numerous young cod about ten inches in length were captured on the hard ground near Crail this autumn[1]. The comparative regularity of their size, indeed, attracted notice, and demonstrated how other methods of fishing than trawling could decimate the young fishes. As formerly mentioned, such forms were caught simply because they held the ground where the lines were "shot." If a trawl had passed through the area many would likewise have been captured.

The grey gurnard had a total of 1,059 or 14 per haul, an average of 5 over last year. Of this number 118 were unsaleable, but of course at certain ports all gurnards are generally unsaleable. This species also shows the zigzag records of a food-fish throughout the ten years. Beginning with an average of 10, it twice fell below this, viz. in 1890 and 1894, when the average was 9, twice it stood at 13, viz. in 1888 and 1889, three times at 15, viz. 1887, 1891, and 1893, and only once reached 19, viz. in 1892. No species could more clearly illustrate the futility of attempting to increase the supply of fishes in the open sea and its neighbourhood by the closure of inshore areas. The numbers remained very much at the end as they were at the beginning, and throughout showed only the ordinary variations. This species seems to reach its maximum in May and June in the area of the Forth, and to diminish in the winter months. It is also in considerable abundance in July, August and September. In the Review in the 14th Report of the Fishery Board it is pointed out[2] that gurnards increased by an average of 3 per haul during the last five years of the experiments, and it is curious that the number of hauls of the trawl during May was greatly in excess of those in the first period. Thus 18 hauls during May in the first period were balanced by 44 in the second. Further, of two years in the latter

1898, vide p. 30. [2] p. 141.

period the number of hauls in May was the highest in the whole series of months in the ten years, viz. 14 in 1891 and 13 in 1892, and producing respectively 597 gurnards in 1891 and 730 in 1892. These alone would make a considerable difference in the average of the second period of five years; indeed, the figures to be pitted against them during the whole of the first period in May are just 181 and 139 respectively. The gurnard, therefore, was under different circumstances in the second half of the experiments.

In glancing at the monthly aggregates during the ten years in the Forth, the effect of the seasonal variations just alluded to in the gurnard are striking. To have drawn conclusions from a series of experiments carried out in January, February, March, November and December would have given very erroneous results, for the total of the gurnards captured in these months amounted to only 93, not a fifth of those secured in April or in October, while these five months give numbers only a twenty-eighth of those secured in May, the month in which the maximum occurred. As May is the month in which the majority of the gurnards are ripe, this increase is apparently associated with the function of reproduction. It may be questioned also whether, in view of this natural consensus in the warmer months, any efforts made by man to alter the balance would have much effect, since the forces at work are beyond his control.

The total number of lemon-dabs in 1895 was 1018 or an average of 13 per haul, a number less by three than in 1886, when the experiments commenced, though one more than in 1894. The reviewer in the 14th Report of the Fishery Board accordingly emphasises the fact that the average of this valuable food-fish had fallen both in the closed and in the open areas during the last period of five years, a result probably due to depletion of the offshore grounds. The first period, however, cannot be safely contrasted with the second in regard to this species, the maximum captures of which occur in June, July, August and September, since the numerous blanks in its colder months exaggerate the captures during its warmer months, more especially if these also are marked by unusually high numbers. Nothing, however, must be taken on trust in so important an inquiry and accordingly a Table has been

Lemon-Dab, during the months of the Decade.

	1st period			2nd period			Average	
	No. of Hauls	Sale-able	Unsale-able	Sale-able	Unsale-able	No. of Hauls	1st period	2nd period
Jan.	9	10	7	110	13	27	1	4
Feb.	18	20	—	150	9	44	1	3
Mar.	18	63	—	147	23	32	3	5
Apr.	18	156	—	339	62	36	8	11
May	18	246	—	505	105	44	13	13
June	35	713	151	471	98	37	24	15
July	16	211	7	668	157	34	13	24
Aug.	35	801	261	794	105	38	30	23
Sept.	24	682	104	624	164	35	32	22
Oct.	31	298	13	377	28	35	10	11
Nov.	29	176	63	138	14	29	8	5
Dec.	18	53	3	131	16	41	3	3
							146	139

constructed showing the captures of the species throughout each month of the ten years. From this it will be seen that the averages for the first period of five years exceed by 7[1] those of the second period, a result in keeping with the principles already so often adverted to, though not so strikingly shown as in certain cases. This result is due not to any noteworthy diminution of the species, but to causes connected with the periods of capture and the usual variations met with in all methods of fishing in the sea. It would have been unusual to find that with so favourable a first period of five years the captures of lemon-dabs had not exceeded those of the second period. This result is, perhaps, more evidently exhibited in the monthly totals for the decade in Table on p. 174. The spindle formed by the captures reaches its maximum (with a slight break in July) in August (1,961), the minimum being in January (133). No fish, indeed, could more conspicuously show the effect of the season in regard to its numbers, and this altogether irrespective of man's interference. It would seem, therefore, that the hypothesis of the depletion of the offshore grounds

[1] Decimals are here as elsewhere not counted.

FORTH.

Totals for Ten years 1886–1895.
(Saleable and Unsaleable.)

	Jan.	Feb.	Mar.	Apr.	May	June	July	Aug.	Sept.	Oct.	Nov.	Dec.
Haddock	729	1888	545	1540	1017	4325	6175	11190	6959	6194	1572	1660
Whiting	947	574	464	696	841	836	1595	2721	2298	3633	1395	1421
Cod	990	955	659	500	446	634	666	1047	842	909	858	1028
Gurnard	1	1	15	545	2676	2145	1223	1540	1309	552	65	11
Wolf-fish	28	52	57	75	88	54	12	6	1	2	3	9
Plaice	928	1901	1425	1608	1982	2400	2844	4465	3426	1673	1579	922
Dab	321	498	765	1142	1483	2449	2643	4678	4740	2535	1233	766
Long-rough Dab	730	1117	887	769	1269	1304	1285	2062	2163	1749	1160	1301
Lemon-Dab	133	179	233	557	856	1433	1043	1961	1574	736	391	203
Craig-Fluke	14	45	59	42	94	58	61	56	76	89	48	32
Flounder	5	24	116	103	34	39	9	4	11	1	3	9
Turbot	2	2	3	4	6	3	4	11	9	4	22	2
Brill	4	4	8	3	1	3	2	4	8	7	7	7
Sole	—	3	—	—	1	54*	98*	1	4	—	—	1
Grey Skate	11	29	20	19	20	38	21	61	76	49	28	18
Thornback	41	47	52	44	56	44	59	107	71	101	74	48
Starry Ray	15	27	12	21	7	20	17	16	10	19	15	29
Frog-fish	26	43	57	69	122	77	115	156	153	106	95	95

* An error apparently exists here, in the year 1886, as the numbers seem to be too large for the sole. The figures 54 in June and 98 in July probably refer to the lemon-dab usually called "sole."

gains no support from the work of the "Garland" in connection with the lemon-dab.

In relation to the other forms captured in the area of the Forth during the ten years the following facts suggest themselves. Flounders remained at the end of the period very much what they were at the beginning, and as it is not a fish specially sought by any class of fishermen, and is one not affected by the offshore supply of eggs, it might have responded to the view that the closure increases the number of food-fishes within the limits. No such tendency, however, is seen either in this species or in the halibut, turbot, brill, sail-fluke sole or topknot. All the latter appear as accidental captures and in small numbers. In the monthly aggregates during the decade the maximum for flounders occurred in March, followed closely by April, these two months, especially March, being associated with the spawning-period. The witches or pole-dabs occur more plentifully than the foregoing, show similar yearly variations in numbers, and were not in any visible way affected by the ten years' operations. The monthly totals for the craig-fluke (witch) are, like those of the flounder, comparatively small, the maximum being in April, and the next in October. There is considerable irregularity in the series, though the colder months have the smaller numbers. The same remarks apply to the hake, ling, conger, poor cod, bib, mackerel, pollack, John-dory and sea-trout. Some of these, as the sea-trout and John-dory, were only met with on one occasion, and the personal experience of nearly fifty years of the neighbourhood shows that this is in accordance with former conditions. It is well known that both salmon and sea-trout are common in the sea, and not always close inshore, yet how seldom are they captured in a trawl. Their wide distribution is only now and then disclosed by the rare capture of a young sea-trout in a herring net. Other fishes like the gar-pike (*Belone*) and skipper (*Scombresox*) are similarly situated, for while it is known they occur in considerable numbers, as the stomachs of sharks and other fishes show, they are captured as a rule by neither liners nor trawlers.

Four round fishes yet remain for consideration, viz. the angler, wolf-fish (cat-fish), herring, and sprat, and it is interesting that two of these have pelagic eggs (for we may include the large eggs of the angler carried in sheets of mucus in this category for the moment), and two demersal or attached, viz. the wolf-fish and herring. The frog-fish (angler), so far as the

The Angler or Frog-fish, a destroyer of food-fishes.

experiments show, was not affected by the closure either in regard to increase or diminution, and the few sprats captured indicate, in waters abounding with such small forms, how difficult it is to estimate their numbers. The pelagic herring, again, is quite independent of such measures as the closure, and the few captured only indicate the shoals that frequent the mouth of the Forth, especially in the first four months of the year. Judging from the captures of the "Garland," the autumn herrings are much less numerous. Thus of saleable herrings in the ten years 45 were captured in January, 49 in February, 8 in March and 16 in April, while a blank occurred till November and December, when 2 and 11 respectively were obtained. Herrings under 7 inches were captured in the other months of the year (with the exception of August) besides those mentioned, as shown in the accompanying table. February seemed to be the month in which the maximum number was obtained.

In considering general fishery questions the history of the herring offers interesting reflections. In our youth the herrings were caught as a rule from 12 to 16 miles from shore, and vast multitudes in August at the Old Hake, a region off the Fife coast between Barbet Ness and the Carr Rock, a distance of about 3½ miles[1]. Shooting their nets close inshore over rocks and shingle, it used to be a common saying amongst the men "no herring no nets," since the plentiful captures of herrings buoyed their nets from the sharp rocks, and *vice versâ* the empty nets were ruptured by contact with the rocks. Since that date the herrings have deserted the Old Hake, so that for many years no fishing of note occurred there, and even now it is only occasionally that they return. Viewing the question broadly, it is generally admitted that the boats require to go further out to sea for the capture of herrings than in the olden time. Yet few would assert that this is a proof of the decadence of the herring-fishery and the diminution of the species. Applying the same argument to the haddock, it is remarkable how promptly the same sequence is brought forward as a proof of its approaching extinction at the hands of the ubiquitous trawlers. The mere fact that fishing has now, even if it were always the case, to be carried on at a greater distance from shore than in the days of our forefathers does not indicate the ruin of a fishery, especially when other features are considered. Again, the annual consumption of whitebait has no appreciable effect on the abundance of the herring.

On board the large sea-going trawlers this fish is a rarity, only 14 having been procured in January and February and 1 in August during the Trawling Expeditions of 1884, chiefly in the neighbourhood of the Forth. Such captures are in a sense accidental since the fish as a rule can readily escape through the meshes.

[1] It is noteworthy how few herrings and sprats occur in the trawl, and such are no test of their abundance in the area, *e.g.* St Andrews Bay. Nets may even be shot from the rocks and capture many, while they are altogether absent from the trawl.

It need not be said that they give but a feeble indication of the vast numbers of herrings occasionally present in the inshore and the offshore waters, yet they aid in familiarizing us with the true facts of the case.

The Forth is a favourite habitat of the cat-fish or wolf-fish, which is "born and bred" within its limits, especially on the stony ground near Crail. In no form is the contrast greater between the Forth and St Andrews Bay, since they are seldom caught in the latter. The cat-fish, moreover, from its habits, is one that ought to have been benefited by

	Herring			Wolf-fish (Cat-fish)		
	Unsaleable	Saleable	Total	Unsaleable	Saleable	Total
January	7	45	52	—	28	28
February	51	49	106	2	48	50
March	8	8	16	3	54	57
April	28	16	44	1	74	75
May	10	—	10	1	87	88
June	44	—	44	1	53	54
July	10	—	10	—	12	12
August	—	—	—	—	6	6
September	1	—	1	—	1	1
October	2	—	2	—	2	2
November	15	2	17	—	3	3
December	11	11	22	—	9	9

Wolf-fish.

1886	1887	1888	1889	1890	1891	1892	1893	1894	1895	Total
—	—	—	1	—	5	7	4	5	6	28
—	—	—	7	3	12	2	10	13	5	50
—	—	—	1	6	4	26	—	16	4	57
—	—	—	12	7	7	9	10	19	11	75
—	—	—	6	10	21	4	16	—	31	88
—	1	4	7	—	—	32	7	3	—	54
—	—	—	2	1	2	1	2	—	4	12
—	—	2	1	—	2	—	—	—	1	6
—	—	—	—	—	—	—	—	1	—	1
—	—	—	—	—	—	—	—	—	2	2
—	—	1	1	1	—	—	—	—	—	3
—	—	—	—	1	5	—	2	—	1	9
—	1	7	38	29	58	81	51	57	65	

the closure. It is a comparatively sedentary form in adult life, being only pelagic in its post-larval condition. But it is a species frequently caught by the liners, and hence was affected by other agencies. Beginning with a single example in 24 hauls in 1887, the average slightly rose in 1888, and still more,

Mending herring-nets after a successful season, July, 1896.

To face p. 178]

in an irregular manner, in the subsequent years, never, however, reaching the average of 1 per haul. The months in which the highest numbers occur are the first six of the year, the captures increasing from January to May, and then falling to June, though the latter is almost double that of January. The two highest figures occur in April and May, viz. 75 and 88, followed by 57, 54, 50, 28, the numbers respectively for March, June, February and January. The contrast with the second six months of the year is sufficiently bold, July has 12 and the others in succession 6, 1, 2, 3, 9. Perhaps no great weight need at present be placed on this feature, but it is evident that the early months of the year were the most productive in wolf-fishes, and it is these months which are wholly absent in the first period previous to 1889. The highest monthly numbers certainly occur in the second half of the period, but the numbers are small, at most an average of 3 per haul. The circumstances of the two periods are assuredly not equal, but it would only be fair to say that this species seemed to be benefited by the closure. One of the most interesting facts, however, is the prevalence of the species during the first six months of the year in contrast with its paucity in the remaining months.

Of the Elasmobranchs, or cartilaginous fishes, the numbers were comparatively small. The thornback had its highest

FORTH.

Monthly Totals, 1886–1895.

	Thornback			Grey Skate			Starry Ray		
	Sale-able	Unsale-able	Total	Sale-able	Unsale-able	Total	Sale-able	Unsale-able	Total
Jan.	24	17	41	7	4	11	6	9	15
Feb.	32	15	47	21	8	29	7	20	27
Mar.	32	20	52	15	5	20	4	8	12
Apr.	42	2	44	19	—	19	10	11	21
May	55	1	56	16	4	20	3	4	7
June	42	2	44	35	3	38	5	15	20
July	55	4	59	18	3	21	8	9	17
Aug.	101	7	107	44	17	61	8	8	16
Sept.	65	6	71	48	28	76	3	7	10
Oct.	91	10	101	40	9	49	5	14	19
Nov.	56	18	74	24	4	28	8	7	15
Dec.	34	14	48	15	3	18	15	14	29

numbers in August and October (107 and 101 respectively for the months of the decade). In the totals of the months nothing striking occurs. The smallest numbers (41 and 44) are in January and April. No evident interference with the ordinary operations of nature in this species seems to have followed the closure. The same may be predicated of the grey skate, which had its highest monthly numbers in August and September

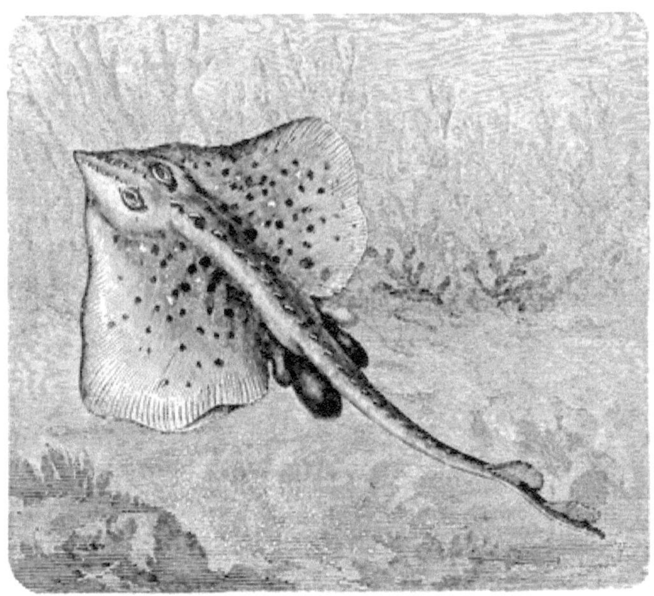

Skate gliding through the water.

(61 and 76), and the lowest in January and December (11 and 18). The monthly totals of the starry ray have no apparent regularity in their increase and diminution. The highest numbers occur in December (29) and February (27). The others were captured too rarely to be of service. A consideration of this group affords no grounds for believing that the food-fishes of the Frith of Forth have been in any noteworthy

manner altered by the closure. Further, while there is a tendency to an increase in the warmer months, the differentiation of the colder from the warmer months is less pronounced in this group than in the bony fishes. The numbers, however, were so small as to make it hazardous to rely on them. The notion that young skate probably frequent rocky ground and are thus free from injury by the trawl is unlikely. They occur on the smooth ground where their parents live, as in the case of the thornback in St Andrews Bay.

Special attention was drawn in the Trawling Report of 1884 to the necessity for observing the increase in the size of inshore flat fishes during the experiments after closure of a bay[1]. It was commonly believed and generally stated by fishermen that the flat fishes (pleuronectids) in open sandy bays had been reduced in size by trawling, that is, all the larger forms had been swept off. There were reasons, however, in 1884 for desiring further information on this point, and hence it was one to which the attention of the Fishery Board in carrying out the experiments was specially directed. One of these reasons was that in the deeper offshore water only adult plaice had as a rule been found, whereas in St Andrews Bay and similar regions an immense number of plaice of small size occurred. Even at that time (1884) an attempt was made to locate the several sizes of plaice and other flat fishes by trawling in parallel lines from the region beyond low-water mark outward. The remarkable uniformity of the saleable plaice (7—13 inches) procured in such a bay as that of St Andrews was another fact which suggested the true explanation, viz., that the eggs, larval and post-larval plaice sought the shore, and that the older plaice as they grew large sought the deeper waters. Year after year since that date and the closure the same size of saleable plaice from St Andrews Bay has uniformly been obtained, showing that the original view was correct, and that little alteration of any note had occurred from the interference of man by the closure.

In the summary in the 14th Annual Report it is stated that from a "calculation of the average size of the individuals

[1] *Op. cit.* p. 379.

of each species in each haul[1]" it is apparent that a small yet definite and appreciable diminution in size has occurred in the area under consideration. This laborious calculation shows how anxious the members of the department were to sift the case as far as they could. But it seems to me that another and more simple solution of the question may be found. If we compare the immature (unsaleable, that is, small and very small) fishes captured in the two periods of five years, it appears that 12,551 occurred in the first period, or 46 per haul, whereas in the second period 33,142, or 74 per haul, were obtained. The larger number of small fishes secured in the second period, when work was much more thoroughly carried on (447 hauls in contrast with 270), would of itself have a tendency to lower the average in each species, and make the reduction to which the reviewer alludes. It is, however, stated under the cod, that about equal numbers were measured in the two periods; but this affords little information, since it is not explained whether the larger or the smaller series were taken in the greatly preponderating second period. Until these doubtful points are made clear it is necessary to withhold sanction from the statement that the size of the food-fishes in the closed waters has diminished. So far as personal observations have gone, the food-fishes in the areas have been throughout the ten years very much as they were before the closure.

In the review of the experiments of the "Garland," in the Fourteenth Annual Report of the Fishery Board, great value is attached, just as in the case of St Andrews Bay, to a comparison of the work of the first five years with the work during the second five years in the Forth. A preliminary comparison is made of the captures in the closed area in 1887 and 1889, the report showing that the average number per haul was more than double in the former what it was in the latter. This gives a key to the whole position, for any conclusion drawn from a comparison of 23 hauls in the months of June, August, and September in 1887 with 90 hauls scattered over ten months in 1889 is hopeless, especially when it is found that five winter months with their low averages form half the period. It is

[1] *Op. cit.* p. 142.

immaterial whether we take the closed or open areas separately, as in the Report, or group them together, as in the present case (Tables XI. and XII.). In the comparison of the captures of the flat fishes in the closed and open areas the same misapprehension is made, the difference of 100 per haul readily falling to 1887 with its working period confined to the three productive summer months. In like manner the notion that on the cessation of trawling there appears to have been an increase in the abundance of fishes within the closed areas, "as shown by the high averages in the year 1887[1]" must lapse. The choice of 1892 for a low average of flat fishes in St Andrews Bay is in the same category, for while in that year 16 hauls occurred in the moderately productive months of May, June, and October (July, August, and September with their high figures being blank), no less than 28 hauls were in the colder months of February, March, April, November, and December. Fluctuations in these and similar cases are due to other causes than a dearth of fishes. In the Report, however, it is pointed out that the curve of fluctuation in the abundance of round and flat fishes corresponds with the general curve for the food-fishes, a conclusion which is in accordance with the method adopted here, viz., dealing with the subject as a whole. From the figures in the Report it would appear that in contrasting the two periods of five years there was an increase of fishes of all kinds in the closed area of the Forth from 242 to 252, or 10 per haul; moreover, that there was an increase in the open area from 160 to 171, or 11 per haul. In this total the flat fishes showed very little change, a diminution of 2 in the closed and 1 in the open only occurring between the first five years and the second. But, on a consideration of the circumstances attending the work of the two periods, it would seem that this diminution might well have been larger, since of a total of 269 hauls no less than 159 occurred in the warmer months (May to October), as against 110 in the colder. Fifty hauls in favour of the warmer period is a very considerable advantage, and it would not be unfair to conclude that if the proportions of the warmer and colder periods in the first five years had

[1] 14*th Report S. F. B.* p. 148.

been nearer each other, there would have been an increase instead of a diminution, as in the case of fishes of all kinds in the Forth. The conclusion[1] therefore "that there has been a decrease in the abundance of flat fishes in the closed waters of the Forth" rests on the most uncertain foundations, and is of little moment in dealing with so large a question, especially when there is appended to it the theory that the cause is the over-fishing of the offshore waters in the neighbourhood.

In the second period (1891—1895) the hauls were grouped as follows, viz., 214 in the warmer months and 219 in the colder months, or 5 in favour of the colder months. The very large number of hauls in the warmer months probably gave a total so substantial that not even the numerous hauls in the colder months could reduce the average, especially since round fishes were a prominent feature of the area, and occasionally occurred in these colder months in considerable numbers. If the three very productive months of July, August, and September in each period are taken, it is found that in the first period (1886—1890) there were 16 hauls in July, 35 in August, and 24 in September, or a total of 75 hauls, whereas in the second period 34 occurred in July, 38 in August, and 35 in September, making a total of 107. Out of the six colder months only one blank period (in November) occurred, whereas there were five blanks in the warmer months. The fact, however, that there was no falling off under these more equally balanced circumstances of warmer and colder months argues well for the food-fishes of the Forth, and gives no ground for despairing of the situation.

[1] *14th Report S. F. B.* p. 138.

CHAPTER V.

SCIENTIFIC INVESTIGATIONS IN THE MORAY FRITH.

TABLES XIII. TO XXVI. 1887—1897.

IN 1884 the Moray Frith was worked only by some of the comparatively few trawlers then in Scottish waters, but it was not searched with that avidity which marked the pursuit of fishes within its area a few years later. Few opportunities occurred in 1884 for its investigation, and, besides, it did not form so prominent a subject in the fishery affairs of the day as it does now. Fortunately, however, in April 1884 three hauls of the trawl were made, in the ordinary course of work, off the coast of Caithness and on Smith Bank, and these afford an important comparison with the subsequent work of the "Garland," and with the region just outside the limit as observed in 1898.

The remarks made on this area in 1884 were as follow:—"The three hauls on Smith Bank (8th and 9th April) afford an interesting comparison with the Bay of Aberdeen. Thus they gave more cod than the 15 hauls in the latter (37 against 38), and also an excess of young cod (codling), viz. 21 against 16 at Aberdeen. It cannot be said, therefore, that the adult cod came to Smith Bank for the purpose of spawning. In their stomachs were young haddocks, young whiting, and sand-eels, so that the latter occasionally go to the deep water at a distance from land. The large size of the haddocks is noteworthy, for none of the kind termed small were procured. The average number per haul was 604, or next to the Forth in rank in this respect. The paucity of whiting was equally noteworthy, while grey gurnards especially abounded, almost all being full-grown fishes, only one or two from 7 to 8 inches having been obtained. The proportion per haul is 116, whereas in the Forth it is 59. The cat-fishes were also numerous, but being unsaleable at Macduff were thrown overboard. The vessels did not seem to work much longer on this bank, a considerable diminution of the fishes having occurred. The captures made on the present occasion were not heavy, and the long journey was expensive."

The foregoing is probably a reliable view of the condition of the fishes of the region at a time when little agitation existed in the neighbourhood, and when the strong statements concerning the depopulation of the waters of the Frith by trawlers were seldom heard.

The three hauls of the trawl produced 2,711 saleable fishes, or 903 per haul (Tables XIII. and XIV.), a very moderate number in contrast with some of those off the Forth the same season, and which produced from 1,500 to 2,744 per haul of saleable fishes. Indeed, the oft-quoted St Andrews Bay occasionally gave as many saleable fishes per haul in 1884, so that in those early days when trawling had made comparatively little progress in the Moray Frith the captures were not remarkable. Only a single grey skate and a single starry ray occurred. The total number of cod was 102, or 34 per haul, a considerably higher average than generally met with during the experiments, yet it was not to be compared in value with three hauls off the Forth 9 or 10 days later in the same season, since the majority (64) in the Moray Frith were young cod (codling), only 38 being adults, whereas off the Forth three hauls produced 84 adult cod and only 13 codling. In the latter area a congregation had probably taken place at the spawning season, whereas in the former (viz. the Moray Frith) such a congregation may only have been commencing. All the haddocks captured were large, and their numbers were considerable, viz. a total of 1,812, or 604 per haul; yet this fell far short of such a haul as II. (off the coast of Haddington) in January, which gave 959 large and 905 small haddocks. The number, however, far exceeds that captured in April, 1884, in any other locality[1]. Whitings were much less numerous than off the Forth, though large, and ling were more frequent. Plaice were in considerable numbers and large, viz. 64 per haul, and the majority had yet to complete spawning. Dabs were not numerous, only 14 being present in each haul. Lemon-dabs were common, no less than an average of 50 occurring in each haul. Pole-dabs or "witches" were very few in this part of the Frith. Gurnards were plentiful and all of a saleable size, the average per haul being 116, a high figure. The wolf-fishes were also in considerable numbers, viz. 9 per haul.

The main features, therefore, of the fish-fauna of the region were the numbers of large haddocks, cod, plaice, and gurnards. These data are reliable for the season, and are in a different category from the vague and unsatisfactory, though reiterated remarks often produced for comparison of a particular area with the conditions at the present day. Moreover, it is well-known that the trawling-ships eagerly frequented the Moray Frith to the last moment, and that to-day all would return if it were possible.

The work of the "Garland" in the Moray Frith commenced in August 1887, fully three years after the preceding observations. Six stations were chosen close inshore as follows:—

I. Off Burghead Bay, length 7 miles, 6—7 fathoms, sand.

II. Off Nairn, length 8 miles, 9—12 fathoms, mud.

III. In Cromarty Frith, length 4 miles, 3—7 fathoms, sand.

IV. This and the succeeding lie in Dornoch Frith. Parallel to shore for 5 miles, 7—9 fathoms, mud and sand.

V. Further out, length 8 miles, 11—15 fathoms, mud and sand.

VI. South of V., length 5 miles, 10—15 fathoms, sand.

Each station was trawled over once in the latter part of August, and the total result was 1,567 fishes, or 261 per haul. Of this number 600, or 100 per haul, were saleable, and 967, or

[1] The note of 8th April, 1884, states that a few of the haddocks might have been counted among the "small" of the Forth, but at the same time they would have been indifferently placed in either group. This shows that the haddocks here are more uniformly large. There are, however, comparatively few very large.

161 per haul, were unsaleable, a comparatively poor result when contrasted with the work either in St Andrews Bay or in the Forth during the same month. Further, the contrast with the three hauls of the converted tug "Royal Saxon" (a paddle steamer) in April 1884 is still more marked, for the average per haul was no less than 903, and all saleable, and it must be remembered that August is the most prolific month of the year. Much of the divergence is probably due to the inefficient work of the "Garland," and the position of the stations.

Plaice held the first position with a total of 559, or 93 per haul, a comparatively high average, though much under that obtained in St Andrews Bay during the warmer months, and even during the entire year, as in 1890, when the average per haul was 169: comparatively few were large plaice.

Dabs followed, with a total of 370, or 61 per haul, an average requiring no comment, further than that it was moderate, though larger by far than the captures at Smith Bank in 1884, or beyond the Frith in 1898. The total for haddocks was 318, or 53 per haul, a small number, and in marked contrast with the captures in April 1884 in the Frith, and in 1898 beyond it; the former showing an average of 604 per haul, and the latter of 695, in both cases the numbers being more than ten times higher. Gurnards were fairly numerous, the total being 182, or 30 per haul. Lemon-dabs averaged about 7 per haul, and thus were more numerous than in St Andrews Bay. It is doubtful, however, how much reliance should be placed on the "Garland's" work, since in the commercial trawlers the average in 1884 was 50, and in 1898 (outside the Frith), 21. Long-rough dabs were very few, about 2 per haul, "witches" (pole-dabs) about 3 per haul, flounders and grey skate about 2 per haul, turbot and brill about 1 in three hauls, cod 5 in 6 hauls, and whiting about 3 per haul. The captures on this occasion showed that fishes in all probability were fairly numerous, and only required efficient apparatus and persistent energy to make their pursuit highly profitable.

The most productive stations were II., V., and VI. Station III. in Cromarty Frith seemed to have slender capabilities.

In 1888, the month during which the six stations in the

Moray Frith were examined was May, and though the number of hauls was the same as in 1887, a considerable diminution was to be expected. The total number was 1,158, or 193 per haul, and the proportions of saleable and unsaleable were widely different from last year, being for the former 1,108, or 184 per haul, and for the latter 50, or 8 per haul.

The gurnard headed the list with a total of 235, a somewhat unusual position for this species, which is generally a subordinate one. In May, however, their numbers are as a rule noteworthy. The average per haul is 39, or 9 over that of the previous year. Dabs follow with a total of 201, or 33 per haul, only a little more than half the number captured last year. Haddocks are in the same category, with 23 per haul instead of 53. The important plaice was represented only by a sixth of the number of 1887. Whitings, on the other hand, were six times as numerous, and witches nearly seven times as plentiful. Long-rough dabs had increased even more largely. The numbers of each species were, however, small, and the changes were due not only to season but to the variations inseparable from such operations, especially in a ship like the "Garland."

A great increase on the previous year occurred at Station I. and a less increase at Station V. All the rest showed a considerable decrease. On the whole the captures were small.

The six stations were examined once in June in 1889, the result being the small total of 634 fishes, or 105 per haul, not half the number with which the experiments commenced in 1887. Of this number 546, or 91 per haul, were saleable, and 88, or 14 per haul, unsaleable. The closure certainly did not seem to improve the captures of the "Garland," and in the Report of the Fishery Board it is said that the results "show a decrease both in round fish and in flat fish."

Though plaice were three times as numerous as in 1888, the total gave only 47 per haul, a moderate number. The figures in almost all the other fishes were reduced. Altogether the year was an unproductive one in the Moray Frith.

The year 1890 was characterised by an increase in the number of hauls in the Frith to three times the former number, and in three separate months, viz., May, July, and September.

The total number of fishes was 2,421, or 134 per haul. Of this 1,433, or 79 per haul, were saleable, and 988, or 55 per haul, were unsaleable. The year showed therefore an improvement on the preceding, though the captures were still comparatively small.

Dabs held the first place with a total of 1,337, or 72 per haul, an average only once reached in the Frith of Forth, but which had been doubled in St Andrews Bay. Plaice came next with an average of 30, a number more characteristic of the Frith of Forth than St Andrews Bay, where the average is high. The other fishes were in small numbers, and indeed the three examinations were not productive. For instance 17 hauls in the much-trawled St Andrews Bay gave in 1886, the year in which the closure was applied, 555 more fishes than the 18 did in the Moray Frith. In the Ninth Report of the Fishery Board it is stated that there was "an increase of flat fish and a slight decrease in round fish."

An increase on the previous year occurred at Stations II., IV., V. and VI., and a diminution at I. and III.

Six hauls in the month of September 1891 gave a total of 1,385 fishes, or an average of 232 per haul, a considerable improvement on 1890. Of this total 875, or 145 per haul, were saleable, and 510, or 85 per haul, unsaleable.

The increase on the previous year was mainly due to a large proportion of dabs, viz. 141 per haul, and an increase of 7 per haul of plaice. Haddocks had reached the average of 3 per haul, the lowest in the series up to date, showing that they at least either did not frequent the areas or escaped the "Garland's" trawl. Gurnards had increased 10 per haul on the previous year, and there were eight hake.

All the stations (especially II., IV., V.) showed an increase except I., where the reduction was nearly one-half on the previous year.

Six hauls in September in 1892 produced a total of 1,745 fishes, or 290 per haul. Of this 1,043, or 173 per haul, were saleable, and 702, or 117 per haul, unsaleable.

While in the Board's Report of the year the concluding sentence of the summary is to the effect that, "It is a noteworthy circumstance that although the prohibition of beam-trawling in territorial waters must have served to protect immature plaice (owing to their very special distribution) more than the young of other fishes, this fish is diminishing in abundance year by year," it is curious that in the Moray Frith, the records of the year show that the number of plaice captured there was nearly five times that of 1891, and the highest in the whole decade. Such are the uncertainties of fishing operations. Each haul yielded 174 plaice. Dabs, on the other hand, were reduced by nearly one-half.

The phenomenal total (for the region) of 1,238[1] on Station IV. along the shores of the Dornoch Frith on the 2nd of September, is worthy of note. This was made up of 894 plaice, 309 dabs, and 35 other fishes.

The closure of the entire Moray Frith on the 22nd November, 1892, caused the Fishery Board to increase the number of stations in the area from 6 to 16. The ten additional Stations are as follow:—

VII. 3¾ miles off Tarbet Ness E. by ½ S., length 3½ miles, 22 to 28 fathoms, sand and mud.

VIII. 12 miles off Tarbet Ness E. by S., length 3½ miles, 28 to 31 fathoms, sand and mud.

IX. Parallel with VIII. and 4 miles further east, 3 miles long, 28 to 32 fathoms, sand and mud.

X. 3½ miles from Lossiemouth, 3½ miles long, 13 to 15 fathoms, sand and shells.

XI. Across Smith Bank, 5 miles long, E. by N., 18 to 25 fathoms, sand and shells.

XII. At right angles to XI. on Smith Bank, length 5 miles, depth 18 to 21 fathoms, sand and shells.

XIII. Parallel with XII., outside Smith Bank.

XIV. 3 miles E. by S. ½ S. of Clyth Ness, 5 miles long, 30 fathoms, sand and shells.

XV. 10½ miles S.E. by E. of Ord of Caithness, 6 miles long, 23 to 31 fathoms, sand and mud.

XVI. 18½ miles N. of Macduff, 6 miles long, 34 to 36 fathoms, sand.

[1] To avoid complexity in tables a few (9) unsaleable fishes such as solenettes, pipe-fishes, and sea-scorpions have been included.

These stations extended the points for observation more directly into the Moray Frith.

Thirty-two hauls of the "Garland's" ordinary trawl were made in 1893, and six with a shrimp net. It has been thought advisable to omit the results of the latter, so as to make the comparison uniform, and to separate those of the original stations (I. to VI.) from the new ones (VII. to XVI.).

Twelve hauls on the original six stations gave the best results yet obtained, viz., 3,837 fishes, or 319 per haul. Of this total 2,168, or 180 per haul, were saleable, and 1,669, or 139 per haul, unsaleable. The work was carried on in May and October, two months on the whole favourable for the experiments, though less productive than August. Thus the average (10 years) for May in St Andrews Bay is 251 and for October 223, while for August it is 515. In the same way the average for the Forth in May is 175, October 262, and for August 421. The comparison will show that the result of the work in the Moray Frith in 1893 was exceptionally favourable.

The highest place was occupied by the dab with 1,577, or 131 per haul, yet in this respect it was 10 under 1891. Plaice followed with 1,413, or 117 per haul, 57 under the average of the previous year—closure notwithstanding. Haddocks had not reached so high a figure, 379, or 31 per haul, since 1887. Gurnards were likewise 12 per haul over the previous year. Lemon-dabs were also more than double the number of 1892. The reporter seemed somewhat puzzled[1] by the divergence between May and October, but as the preponderance was chiefly caused by large numbers of small plaice and dabs in the latter month no surprise need be felt. It is observed that the cessation of trawling will, to some extent, account for the increase of flat fishes, and apparently it is thought that the Moray Frith may come up to expectations better than either St Andrews Bay or the Frith of Forth "since it contains within it extensive spawning-grounds for white fishes." Subsequent results, however, seem to demonstrate the value of the saving clause "but it is too soon yet to draw any conclusion as to the results of the closure."

[1] *12th Report S. F. B.*, p. 27.

Twenty hauls took place in May and October on the outer stations (VII. to XVI.), resulting in a total of 4,760 fishes, or 238 per haul. Of this number 2,445, or 122 per haul, were saleable, and 2,315, or 115 per haul, unsaleable.

A considerable divergence in the fish-fauna of these outer stations is at once apparent, and it would have been well if the Blue-book had devoted some of its energies to this feature. Instead of an average of 117 plaice per haul, the average was only 7, but they were all of the larger size, viz., from 12 to 19 inches. The trawl had got rid of the swarms of young plaice, so characteristic of the shallower water. The ubiquitous dab was in less numbers than in the inshore stations, but still held a good position with an average of 69 per haul. Moreover, there was an absence of the line of demarcation as in the plaice, for considerable numbers of the unsaleable and smaller saleable sizes were present. Lemon-dabs were about three times as numerous. Long-rough dabs were more than twice as numerous. One of the greatest contrasts, however, was in the haddocks, which had an average of 31 per haul in the first six stations, and one of 97 in the outer (VII. to XVI.). Cod were also more numerous, and the general distribution of the gurnard was indicated by a higher average (28).

The foregoing captures of the "Garland" in the Moray Frith are, at best, but poor in contrast with the 800 or 900 fishes secured each haul by the commercial trawler.

The same number of hauls (12), as in 1893, on the six inner stations in July and October, 1894, gave a total of 3,018 fishes, or 251 per haul, a result falling short, by 68, of the average per haul of the previous year, one of the usual uncertainties of the pursuit.

Almost all the fishes were reduced in numbers except lemon-dabs, the total for which was double that of the previous year, though still small (74). Plaice were 94 per haul, just one above the condition in 1887, not an encouraging feature in connection with the closure of the "extensive spawning-grounds." Dabs were still comparatively abundant (113 per haul), though 18 under the previous year; their high numbers in comparison with the plaice being interesting in connection with the notion that they and the long-rough dabs supplant the plaice from the facility with which the small breeding individuals escape through the meshes of the trawl, whereas the breeding plaice cannot. The evidence here is not in favour of this view. Haddocks were 10 instead of 31 per haul, and gurnards 13 instead of 22 per haul. The experiments therefore resulted in a moderate capture of fishes.

All the stations showed a reduced number, except IV., where the total was 38 over the previous year, and second in the series of nine years.

If little success attended the work of the "Garland" on the inner stations, it was this year very different on the outer stations (VII. to XVI.), where the same number of hauls (20) as last year yielded 2,392 more fishes, or a total of 7,152, or 357 per haul. Of this number 5,556, or 277 per haul, were saleable, and 1,596, or 79 per haul, unsaleable.

The foregoing increase was chiefly due to the large numbers of haddocks, which amounted to 3,172, or 158 per haul. These were chiefly captured in October, the total for that month on the ten outer stations being 2,204, and they were saleable fishes of 10—14 inches. The contrast with the inner stations is sufficiently marked, for the twelve hauls there in July and October produced a total of only 115 of the same size. The enormous number of haddocks evidently peopling the neighbourhood, and many of which would survive to increase the species, shows how capable nature is to maintain the food-fishes in the open sea. The Moray Frith alone could, in all probability, have supplied in October and the subsequent months of 1894 sufficient haddocks for all the fleets from Aberdeen, and that without causing any undue strain on the permanent abundance of the species. They presented themselves there at that convenient season, yet due advantage was not taken of the opportunity, and the reduced average of the succeeding two years showed that perhaps such opportunities do not occur every year. A favourite argument of some is that while the total for the haddock and other fishes goes on annually increasing, or at any rate maintaining its ground with the usual variations, that this total is made up of fishes brought from Norway, Iceland, and the Faroë Islands—to which distant regions the ships have now to go to keep up the supply. Every Scotch bay, according to these "practical" gentlemen, is "swept out," or "cleaned out," so that it is no longer profitable for liner or trawler to fish there, and the ruin of our fisheries and our fishermen is already accomplished. But, in the first place, they forget that unproductive periods in haddock-fishing occur very frequently, and have occurred from time immemorial, times when the men prepare their boats for herring-fishing (the unfailing nature of which is marvellous), and which often richly rewards their efforts. Moreover, if the ships go to a great distance for large and certain gains in regard to white fishes, it is not because their own waters are depopulated, but because with less effort they secure a rich harvest in other seas. Facts, like those produced by the experiments now under review, are therefore not to be compared with the crude suppositions which serve the purpose of the "fisherman's friend." If by dipping a small trawl like that of the "Garland" in the Moray Frith, 220 saleable haddocks of 10—14 inches in length, and at least 33 others of a small but also of a saleable size, are brought on board each haul, not to speak of other fishes, he would be more than a bold man who holds that our haddock-fisheries are ruined. Other areas and other forms are dealt with elsewhere, but that one fact alone disposes of the idea that any such area as that of the Moray Frith is "swept out" by trawlers.

Dabs still maintained a high average, viz. 87 per haul, or 18 more than the previous year. Lemon-dabs remained the same, while the captures of the long-rough dab had been trebled. Whitings and gurnards were also more numerous, as also were "witches" or pole-dabs, which gave an average of 4 per haul instead of 1 in 1893. Cod, on the other hand, were fewer, about 2 per haul instead of 5 in 1893.

It is curious that all the stations except XI. and XII. show a decided increase. These two stations form a cross over the centre of Smith Bank. It is possible that special influences of season or ship were connected with this condition.

A table was this year given by the Fishery Board showing the quantities of white fishes caught within the closed waters of the Moray Frith, and the number of boats, with a view of noting the changes inaugurated by the closure. Seven districts were chosen as fishing specially within the closed area, but even though such were the case, the difficulties of an accurate record are great, and it may be expected to show only the usual variations from year to year characteristic of all fishing operations. There is small chance of any satisfaction being derived from an accumulation of fishes due to the closure. This table extended up to date (1897) is appended :—

Captures in the Moray Frith by Line-Boats in cwts.

	1894	1895	1896	1897
Cod	52571	4403	64663	79731
Ling	2169	339	3868	3544
Torsk	25	6	94	25
Saithe	6120	552	10636	11761
Haddock	153529	65421	156703	126031
Whiting	5845	3347	4836	3319
Turbot	5	—	15	16
Halibut	254	22	691	707
Lemon-Dab	—	—	19	14
Flounder, Plaice, Brill	5477	208	3402	3978
Conger	1244	49	823	1533
Skate	3281	177	3683	3999
Other kinds white fishes	7976	1967	7483	6663
Total	218495	78491	256916	241350
Average haul or shot	3·5	4·77	4·26	3·83
Large boats	7082	1107	11915	14039
Small boats	54866	14930	48346	48836

The work of the "Garland" on the inner stations (I. to VI.) was carried on in 1895 in July, six hauls of the trawl giving the meagre amount of 715 fishes, or 119 per haul, less than half that of the previous year, a result by no means encouraging with the stake at issue. The vast enclosed area with its "extensive spawning grounds" did not in connection with these stations indicate an improvement. Not even by such a gigantic measure were the arrangements of nature in the open sea to be altered to any noteworthy extent. The saleable fishes were 380 in number, or 63 per haul, and the unsaleable 335, or 55 per haul. A great reduction had occurred in all the fishes usually having the higher numbers, such as plaice, dabs, haddocks, long-rough and lemon-dabs, and gurnards. Every station gave evidence of this feature.

Ten hauls on the outer stations (VII. to XVI.) were somewhat more productive, resulting in the capture of 2,001 fishes, or 200 per haul. This was composed of 1,144, or 114 per haul of saleable, and 857, or 85 per haul, of unsaleable fishes. Dabs were most numerous, viz. 57 per haul, haddocks next at 52 per haul, followed by long-rough dabs and gurnards. At every station except XI. (across Smith Bank) a reduction took place.

The inefficiency of the ship probably had a close connection with the foregoing result.

The year 1896 was characterised by an increase both in the number of months and of hauls of the trawl, for work was carried on in August, October, and November, yet the number of hauls was not great seeing that the ship was free from duties both in the St Andrews Bay and the Forth. It is true the number of hauls was a secondary consideration, for if they had been effective (like those of the commercial trawler) their number would have sufficed.

Eighteen hauls in the three months on the inner stations (I. to VI.) above-mentioned gave a total of 1,553 fishes, or 86 per haul, a result even more depressing than the previous year, for it was 33 per haul less. It has to be remembered, however, that November is one of the unproductive months in the shallower water, and actually furnished only about half the number procured in October. If any reliance was placed by the Board on the work of the "Garland," here certainly the effect of the closure of this great area carried only disappointment. Nor was much comfort to be derived from the statistics of the fishes caught by line in the enclosed waters.

The comparatively cheap dab headed the list with an average of 31 per haul, followed by the plaice at 30 per haul, the important haddock being represented by only 2 per haul, and the gurnard by 5 per haul. The saleable "witch" occurred at an average of a little over 1 per haul, and the thornback at 2 per haul. The four years of the closure of the "extensive spawning grounds" had, so far as these statistics showed, given no sign of success, and if they should be prolonged to 14 or even to 40 years, it is unlikely that the arrangements of nature in the open ocean would be materially altered. Even the laborious collection of the statistics of the white fishes caught by the liners in the enclosed area do not improve the situation, for instead of the 4·77 cwts. per haul of 1895, they show only 4·26 cwts. in 1896. They go far to demonstrate, however, that with an efficient ship the Board's returns might have been different, though such might not have altered the conclusions to be drawn from them.

On the outer stations (VII. to XVI.) the same number of hauls took place—yielding a total of 3,869 fishes, or 214 per haul, an increase of 14 over the previous year. Of this number 2,409, or 133 per haul, were saleable, and 1,460, or 81 per haul, unsaleable[1].

Here also the dabs surpass the others in numbers, a feature seldom or never observed on board a commercial trawler, their average being 94 per haul, or nearly double that of the previous year. The saleable plaice reached only about 2 per haul, lemon-dabs were fewer than in 1895, and so with long-rough dabs and haddocks, which had an average of 43 per haul, in contrast with 2 per haul on the inner stations. Gurnards had increased (chiefly in October) by 5 per haul. One of the features of the Frith, in common with the neighbouring waters, was the frequent occurrence of hake, which are seldom seen in St Andrews Bay or in the Forth. Witches or pole-dabs are also characteristic of the softer grounds.

During 1897 twelve hauls of the trawl took place in the months of June and November (six in each) on the inner stations (I. to VI.), resulting in a total of 1,770 fishes, or 147 per haul, a considerable increase (61 per haul) on the previous year, which was the fourth in a gradually diminishing series. The average is but a moderate one, not half that of 1893, and about half that of 1892. Of saleable fishes there were 1,110,

[1] A few fishes such as the butter-fish, dragonet and pogge are included; but their numbers are insignificant and have no influence on the totals. The angler is only occasionally sold. The unsaleable, therefore, chiefly include the immature.

or 92 per haul, and unsaleable 660, or 55 per haul. While June in other areas has not been one of the most productive months, and November is classed with the unproductive, yet, in this instance, the peculiarity exists of a higher average in November than in June, viz. an average of 205 in the former and less than half, 87, for the latter. A tendency to this condition had been noticed in former years, *e.g.* in 1896, in which the total of 4 hauls in November exceeded that of 6 hauls in August. The marked differences in the captures of haddocks, dabs, and plaice, in the present instance (1897) sufficiently account for this feature. Thus in June they were respectively 10, 196, and 76, whereas in November they were 35, 537, and 473. Plaice thus showed an increase of 397, and dabs of 341. Of the eight months during which operations in the Moray Frith were carried out in the years 1887—1897, a period of eleven years, the highest average occurs in October, September being next, then August, July, November, May, June, and April. November stands thus as fifth in the series. The number of hauls, however, is comparatively small, so that too much weight need not be put on this feature, but it is necessary to allude to it.

As in 1896, the dab was the most plentiful fish, the average per haul (61) being almost twice as great as in the year just mentioned. Only a single dab over 11 inches was obtained, the majority (446) being unsaleable, while 286, from 7 to 11 inches, were saleable. This average (61) was precisely that with which the experiments began in 1887, the numbers falling to half next year, and the following to only 18 per haul, then springing up to 72 and 141, the latter being the highest average during the 11 years, and from work done only in the month of September. It does not follow, however, that this month always held the same position, for the following year the average was only 85. After rising to 131 in 1893, it fell during the next three years to 31, the second lowest average of the series. The average for the five years 1887–91 was 68, and for the six years, 1892–97, 77, but work was carried on thrice in October, the most productive month in this area, whereas in the first period this month was absent. As in other instances the captures of the dab in the Moray Frith presented the usual variations attendant on sea-fishing, and in no sense can it be said that the closure of the area benefited this fish or altered the arrangements of nature.

Plaice followed with an average of 45 per haul, or 15 over that of the previous year. Only 75 were unsaleable, that is, under 7 inches, the majority (249) being between 7 and 11 inches, while 191 were between 12 and 18 inches, only 34 being beyond 19 inches. This average, though higher than the previous year, held a median position in the series, five being below it, the minimum being 14 in 1888, and five above it, the maximum (174) being about four times as great, and occurring in 1892, when the work was confined to the productive month of September. The record of the plaice, after even greater irregularities than the dab, remained in 1897 only half what it was in 1887, and though during the first five years the average was only 40 as contrasted with an average of 73 in the last six years, the conditions under which the work was accomplished in the two periods were divergent. Thus while no colder months are present in the first period, May and June twice appear. In the second period they occur only once, while October is entered thrice. The condition of this species, therefore, leads to the same conclusions as in the dab.

The position of the haddock throughout the eleven years shows great variability, since the highest average (53) occurs in 1887, the first year of the experiments, the last year (1897) having only 3, the same number likewise representing the captures for 1891, the last of the first period. The average for the years 1887—1891 is 14, and that for the years 1892—1897 (six years) being 11, or 3 less. The absence of similarity in the months, however, renders comparison uncertain. Moreover, since even a more marked change is present in these periods in the gurnard, it cannot be said that the condition of the haddock is peculiar. The average for the first five years in the gurnard is 19, for the second period it is 12, that is to say, 4 more than in the case of the haddock. In 1887 the average for the gurnard was 30, in 1897 the average was only 11, yet it would be inconsistent with experience to suppose that these figures indicate a diminution of the gurnard from any method of fishing. The irregularities of the pursuit are further illustrated by the witch, which in the first period was three times as numerous as in the second.

The condition of the lemon-dab in 1897 is interesting, for the number (42) obtained in 12 hauls is exactly what was caught in 6 hauls in 1887. Here then is a case for the oft-repeated assertion that the sea is being depleted, since with double the work only the same number were caught as in 1887. If the months and days had been the same in both periods greater colour might have been lent to this notion, but they are not. A survey of the years shows the ordinary variations. Thus in the first five the highest average is in 1887, and the lowest in 1890, the average for the whole period being 3. In the second period of six years, the highest average was in 1894 and the lowest in 1896, the usual variations also being present throughout the period, which had the same total average, for all practical purposes, as the first period. The long-rough dab stands even in a worse position, for the average per haul for the first period was 6, while in the second it was a little over 3. Would any one of experience maintain that the long-rough dab was yearly becoming scarcer in the Moray Frith? On the other hand, the cod has an average a little under 2 for the first period, whereas for the second it is fully 3. The whiting, again, is the reverse, since it was about 6 times as plentiful in the first five as in the second six years. The hake and the turbot were likewise more numerous in the first period, while the brill was more frequent in the second period.

The witch, a characteristic form of the region, was fully three times as numerous in the first period, and the flounder and the grey skate were more plentiful in the same period. To balance these irregularities, the thornbacks were about twice as numerous in the second period.

It is unnecessary to proceed into the details of these forms as regards months, for the number of hauls was comparatively small. It is sufficient to draw attention to the ceaseless variations, and the danger of drawing conclusions from data that are not thoroughly sifted.

Viewing the captures in connection with the various

stations (I. to VI.), a decided increase is apparent on every one in 1897, its amount being in the order of the stations as follows:—19, 120, 2, 6, 98, 121. Judged by the averages for the first five years, the most productive stations were V. II. and IV., whereas in the second period the order was IV. V. and VI., the highest average of the whole series occurring at IV.

The average for all the stations during the first five years was 170, or 108 saleable and 61 unsaleable, while during the second period (1892—1897) the average was 192, an increase of 22, composed of 122 saleable and 69 unsaleable. The question as to whether this increase was due to the closure or to the nature of the working-periods is best answered by a reference to Table XIX., where it appears that all the hauls in the productive month of October are confined to the second period. It is true one haul in April, with an average of only 35 handicaps this period, but there were 11 hauls in July against 6 in the first period, 7 against 6 in August. Moreover, though 12 hauls occurred in November in the second period, their average was equal to that in May. The advantage of 12 hauls in September in the first period against 6 in the second period could not obliterate the foregoing advantages. The closure, therefore, here as elsewhere, had no evident influence on the area.

During the same months of 1897, 19 hauls of the trawl took place on the outer stations (VII. to XVI.), giving a total of 4,126 fishes, or 217 per haul, an increase of 5 on the previous year. Of this number 2,006, or 105 per haul, were saleable, and 2,120, or 111 per haul, were unsaleable, a smaller number of the former and a larger number of the latter than in 1896. The apportionment of the hauls in the respective years was different. Thus in 1896, only 4 hauls took place in the colder months, while 14 occurred in the warmer, whereas in 1897, 9 were in the former and 10 in the latter. It has to be borne in mind, however, that November sometimes presents in the Moray Frith a considerable average from dabs, gurnards and haddocks.

In the light of the five years' work, 1897 stands third, the highest average, viz. 357, being in 1894. The next is the first year of the experiments on these stations, viz. 1893, with an average of 238, or 21 higher than in 1897. The lowest average occurs in 1895. There is nothing, however, in these totals to indicate any effect of the closure, or other feature than the ordinary irregularities of the pursuit.

The hauls throughout the quinquennial period were fairly distributed over the six months from May to November, except that more than double the number of any other month occurred in October. A very different average would have resulted if such had not been the case, for the average exceeds by 61 that of any other month. It is followed by that of November at 282, only 61 less, then July with 211, May 182, August 176, and June 130. There is thus, so far as the individual months go, a parallelism between the total monthly averages of the inner and the outer stations of the Moray Frith.

The study of the yearly averages of the outer stations gives no sign of accumulation, but tends to corroborate the opinion that man's influence in the way of closure is immaterial.

The fact that the dab holds even a more conspicuous position than in 1896 again emphasizes the nature of the "Garland's" work, which materially differs from that of a commercial trawler. The average per haul is 128 or 34 more than the previous year, and it is the maximum number for the quinquennial period, and is nearly double the average of 1893. The capture of dabs in the Moray Frith seems to be characteristic of October, and as this month held a prominent position, the abundance of dabs is a consequence.

The next average, viz. 26, that of the long-rough dab, is separated by a wide interval. It is more than double that of 1893, yet it is the next lowest in the quinquennial period, and is exactly the average for the latter, viz. 26.

The gurnard follows with an average of 20 per haul, the lowest of the series. The average for the quinquennial period is 28.

This year seems to have been, for the "Garland," an unfortunate one for haddocks, only 17 per haul having been procured, a condition of things probably due to some extent to the months during which operations were carried on, viz. June and November, for the maximum number of haddocks occurs in October[1].

[1] It may be that the handling of the ship in regard to tides and other factors in efficiency had a share in this result.

It is 80 below the average of 1893, and not half what it was the previous year, and no less than 60 below the average for the five years. That this state of matters was not due to absence of the haddocks need hardly be discussed, since, a few months afterwards, a commercial trawler, outside the limit of the Frith, captured 695 per haul[1], and when it is remembered that the liners produced a large number from the enclosed area the same year. Yet it may legitimately be advanced that the averages of the haddocks both in the inner and outer stations have shown a downward tendency from 1887 and 1893 respectively. Thus, in the first five years of the inner stations the average steadily sunk from 53 to 3, while in the second period (six years) it began at 9 and ended with 3. It is true it rose to 31 in 1893, but thereafter it declined to 10, 8, 2, and 3. Now the work was carried on entirely during the warmer months in the first five years, and in the second period (six years) 13 out of the 66 hauls were in the warmer months. The total hauls for the eleven years were 108, and of these 95 were in warmer months, and only 13 in the colder. A survey of the captures in detail according to the months of the two periods gives, however, further information :—

MORAY FRITH.

1887–1897.

		Stations I. to VI.		Stations VI. to XVI.		
No. of Hauls		Total Haddocks	Average per Haul	No. of Hauls	Total Haddocks	Average per Haul
1	Apr.	—	—	—	—	—
17	May	178	10	10	798	79
12	June	71	5	10	200	20
17	July	88	5	11	703	63
13	Aug.	341	26	13	510	39
18	Sept.	107	5	—	—	—
18	Oct.	496	27	30	4215	140
12	Nov.	63	5	13	326	25
108		1344	12	87	6752	77

It is evident that haddocks are less numerous on the inner than on the outer stations, the disparity (12 to 77) being so great as to necessitate different treatment. At the inner

[1] *Vide* p. 207.

stations the maximum period is in October, closely followed by August, then May, the other months being nearly equal. Now the comparatively high average of 53 with which the observations commenced was obtained in August during a year (1887) in which exceptionally high averages were present in the areas of St Andrews Bay and the Forth. Next year (1888), with an average of 23, the work was done only in May, a month which shows an average not half that of August. The third year (1889) had its working period confined to June, in which the average is not half that of May. There was a decrease to 10. The irregularities so common in fishing operations might easily reduce the total average of three "poor" months to four, or one under the lowest, whilst the same cause might readily diminish an average of 5 to 3. No great weight need be placed on the suppositions of the last sentence, yet the whole history of fishing operations is full of such irregularities, which moreover strongly characterise the haddocks in the second period. Thus in 1892 the hauls occurred only in September, yet the result (9) is 4 above the average. In 1893 a rise to 31 per haul, the second highest in the series, took place in the maximum and a medium month, the third (April) being a blank as regards haddocks. Next year, the maximum month and one with a small average only resulted in 10. The last three years of this period had diminishing averages of 8, 2, and 3; yet the first had a working period including one productive month for haddocks, viz. August, whilst the second included both August and October, so that the diminution must have been due to other causes, which also seem to have affected the outer stations since their average was smaller during these years (*vide* Table XXVI.). Haddocks apparently were not plentiful on the inner stations during the working periods, but their numbers increased seawards, as the returns from the outer stations and the captures of the liners in the closed area prove. The haddock, indeed, was by far the most prominent fish in the returns from that region. Thus in 1894, 153,529 cwts., in 1895, 178,370 cwts., in 1896, 156,703 cwts., and in 1897, 126,031 cwts. of haddocks at least came from the closed area[1], independently

[1] *16th Report S. F. B.*, Part III. p. 20.

of the incursions of foreign ships. It has also been proved that fine haddocks were in great numbers beyond the closed area. The diminution, therefore, was local diminution, a feature of perennial occurrence in all seas.

The maximum month, so far as the statistics go, for haddocks on the outer stations (VII. to XVI.), which may now be considered, was also October, and the average for this month is about double that next in order, viz. May, while July, August, November and June follow with diminishing averages[1]. The total average for the outer stations is more than six times that for the inner.

The average for the first year (1893) was 97 haddocks per haul, the work being done in May and October, months having the highest averages for this fish in the returns. Next year a great increase occurred, viz. an average of 158 per haul, the experiments having been carried out in October and July, the latter having an average of 63. Next year (1895) the average fell to about a third of the last-mentioned, viz. 52, the working-periods being in July and August, the latter having an average of 39, and thus the captures correspond with what may be termed the normal. In 1896, a diminution, due to other causes than those connected with the periods, had taken place. A similar reduction is present in the captures by the liners. In 1897, work was carried on in June and November, months having an average of 20 and 25. The actual captures gave an average of 17, the lowest in the series, or 5 under the normal. The returns of the liners also show a diminution.

In respect to the haddock, therefore, there is little to satisfy the country, and especially men of science, as to the propriety of the closure of this great area. If experience, indeed, did not show that variations in areas are common, and that no lack of haddocks existed in hundreds of square miles beyond the area, the Fishery Board, so far as the work of the "Garland" is concerned, is confronted with anything but success in its efforts to increase the products of the sea by controlling the "spawning grounds."

Of the other fishes the lemon-dab was twice as common as

[1] No work was carried on in April and September on the outer stations.

on the inner stations, and remarkably uniform in its average, which was 8 in four years, and 5 in the fifth (1896). Whitings on the other hand were variable and few, and the same may be said for hake, ling, turbot, brill, grey skate, sandy ray and sail-flukes. Flounders were absent. The most frequent skate was the starry ray, so abundant in deep water. Anglers were more numerous than on the inner stations.

In surveying the averages on the several stations (VII. to XVI.) the ordinary variations are evident. Station VII., off Tarbet Ness, maintaining from the first a fairly high average, stands at the head with an average of 392 per haul for all kinds of fishes. Station VIII., in the same region, but further seawards, follows with an average of 382, yet in 1894 and 1895 the two highest averages (624 and 614) occur on this ground. The others appear in the following order of averages, IX., XV., XVI., XIV., X., XII., XI., and XIII., the average for which was only 134, or about half that for all the stations, viz. 251. Commencing with a total average of 238 the series concluded with one of 217. The former is made up of 122 saleable and 115 unsaleable, the latter of 105 saleable and 111 unsaleable, that is to say, a diminution in both. Such reduction, however, as the general average shows, is as much due to the usual variations as to any other cause, fishes of all kinds being probably about as numerous at the end as the beginning.

On the 7th and 8th of April, 1898, six hauls in a commercial trawler[1] were made outside the limits of the Moray Frith, resulting in a total of 5286 fishes, or 881 per haul, a contrast to the indifferent work of the "Garland" within it (Tables XV. and XVI.). Of this number 4914, or 819 per haul, were saleable, and 372, or 62 per haul, were unsaleable. It has to be remembered that this work was carried on in the open sea in the midst of numerous trawlers (upwards of twelve in sight, probably from 12 to 20)—all eagerly following the same pursuit. Yet this

[1] For this courtesy the author is indebted to Bailie Pyper, of Aberdeen. An otter-trawl was used. It is a noteworthy fact that, since April, this vessel has been manned by a crew of liners from Torry and Cellardyke, and they are doing profitable work both for themselves and the owners. Moreover, Peterhead, Fraserburg, Hopeman, and Wick give signs of moving trawlwards.

much fished water contrasts most favourably with the three hauls of a commerial trawl[1] made on Smith Bank nearly on the same date in 1884, and resulting in the capture of 2711 fishes, or 903 per haul. In both cases the chief fish was the haddock, 604 per haul occurring on Smith Bank in 1884 and 695 outside the Frith in 1898, a fact which demonstrates that the distribution of this species is so wide, and its numbers so great, that a thoughtful survey of the whole subject leaves little room for doubt as to the wisdom of removing all unnecessary restrictions from fair fishing. The haddocks of 1884 on Smith Bank had fewer small in their ranks than those of 1898 in the open water, but it is probable that a special congregation of spawning fishes had occurred, or, at any rate, had come in the way of the trawl on that occasion. The haddocks of 1898, just beyond the enclosed waters, were more numerous by 91 per haul, and of the total of 4173 only 42 were unmarketable, though some of these would have been eaten at St Andrews. We know that haddocks abounded as of old in the great enclosed area of the Frith, and extended far beyond it in the numbers just indicated, but we fail to observe the slightest trace of any effect on this fish in or near the Moray Frith by the closure of the huge area against British trawlers. Such a step was never dreamt of in the cautious proposal to close the areas of the Forth, St Andrews Bay and Aberdeen Bay *for experimental purposes*,—and yet it has been done without any clear evidence being produced to warrant its adoption. Step by step every available argument has been examined, and its facts duly weighed or its want of facts exposed, but no scientific basis remains on which to uphold such a proposal. A consideration of the condition of the haddock alone in this area in 1884 and in 1898 dispels the fanciful notions of certain of the so-called "practical men," who are responsible for much of the confusion in fisheries' problems still lingering in our country. No imaginary figures or theoretical percentages are dealt with here, but facts resting on personal observation. There is no sign that the extensive fishing-operations in the Moray Frith and its neighbourhood have proved seriously

[1] Beam-trawl of about 51 feet.

detrimental to the distribution of the haddock,—no more indeed than the constant pursuit of the herring has affected its perennial abundance. In the case of both species, periods of scarcity occur, and are used by agitators and others to proclaim the decline, or, it may be, the utter ruin of the fishery, and the urgent need for administrative interference. But both classes are ominously silent when boats are overloaded or nets sunk with their valuable captures, which nature, never failing in resource in the open sea, sends as an answer to their complaints.

To return to the consideration of the examination in 1898, it was found that plaice were much fewer beyond the Frith than on Smith Bank. Instead of 64 per haul, only 3 occurred outside. Some suppose that a congregation of plaice ensues on Smith Bank at the breeding season, but it is more probable that this area is the home of the adult plaice, which is more or less local in its groups, the latter being fed by a constant stream of adolescent forms from the shallower waters. Dabs, on the other hand, gave exactly the same average as on Smith Bank, viz. 14 per haul. Lemon-dabs were not half so numerous in the distant area as on Smith Bank, a feature probably due to their peculiar distribution, unless it be held that the trawl singled out this species and the plaice for special destruction. Long-rough dabs were absent in 1884, but had an average of 6 per haul in the outer waters, a feature in accordance with the view just expressed. Cod were more than twice as numerous on Smith Bank in 1884, viz. 34 per haul, whereas in the open waters in 1898 they had an average of only 14 per haul, but it must not be forgotten that many ships had for weeks scattered their shoals and decimated the stock of the season before these observations were made. Whitings were more numerous in 1898 by 12 per haul. Gurnards, again, were less numerous by 69 per haul. Ling remained about the same; while sail-flukes occurred only in the outer waters to the number of 10 per haul. Witches were also more numerous in the proportion of 11 instead of 1 per haul. One turbot occurred in three hauls in 1884 on Smith Bank and one in 6 hauls in the outer water. Grey skate and starry rays were more numerous in the outer area.

Nothing in the foregoing facts is inconsistent with a sound condition of the area of the Moray Frith (closed and open) as regards food-fishes. They are in unison with what has been found in the other areas, and if they demonstrate anything more than another it is that the interference of man, specially by closure, is powerless to increase the food-fishes of the sea, or by eager fishing to reduce them to vanishing point.

A careful consideration of the returns both of liners and trawlers in connection with the Moray Frith shows that there are no satisfactory grounds for the closure of "spawning areas," any more than for the closure of the inshore limit of three miles, for the purpose of increasing the fish-supply of the country. For eleven years closure of part of the area, and for five years closure of the entire area, has been in force, with the result that, allowing for the usual variations, the fishes in the area and its neighbourhood as a whole, are very much what they were at the beginning. Scientific evidence as to the serious diminution of any species is wanting. On the contrary, there is proof of the abundance of all the important forms over a wide area, including not only the closed region, but the waters beyond. Temporary reduction, local or otherwise, has always occurred, but, sooner or later, the nomad food-fishes again assert themselves, and continue from generation to generation of men a never-failing supply.

It cannot be said, however, that complete satisfaction has been felt with the work as carried out by the "Garland," and if other means for observation had not been constantly at hand, the same confidence would not have been felt in coming to a decision on a question so important. The investigations of 1884, those of the commercial ships before the closure of the Frith, those during its closure, and those made in April, 1898, outside the Frith, together with the work of the "Garland" and other observations, give a body of facts which are reliable.

Lateral view of a Trawler in Aberdeen harbour.

To face p. 210]

CHAPTER VI.

SCIENTIFIC INVESTIGATIONS IN THE FRITH OF CLYDE.

1888—1897.

TABLES XXVII. TO XXXII.

TWELVE stations were selected in the estuary of the Clyde chiefly in relation to the fishing-banks frequented by trawlers, six in the region extending from the eastern side of the Mull of Cantyre to Loch Ryan, three in Kilbrennan Sound, and three between Arran and the Ayrshire coast. Station I. begins a mile and a half south of Davor Island, about 200 yards from the shore, and extends to Davor Lighthouse. Depth 13—17 fathoms; bottom sand. Station II. begins at Plack Point in the Sound of Kilbrennan, about 200 yards from shore, to Carradale Bay. Depth 10—23 fathoms; sand and loose tangles. Station III. in Machig Bay, Arran, and extends 4 miles N. and S. at the same distance from shore as the foregoing. Depth 7—20 fathoms; sand and tangles. Station IV. off Pirn Mill in Kilbrennan Sound, 1½ miles long. Depth 20—29 fathoms; sand and sea-weeds. Station V. is half a mile from shore in Whiting Bay, extending for 3 miles N. towards Holy Island Light. 9—15 fathoms; sand and tangles. Station VI. extends 1 mile east of Rhuad Point, Cantyre, for 3 miles S., and then W. for 1½ miles. 25 fathoms; sand and sea-weeds. Station VII. off Ballantrae, and extends 4 miles parallel to coast. Depth 15—25 fathoms; sandy, with a few weeds. Station VIII. 10 miles off Ayrshire coast, passes W.

14—2

for about 4 miles, ending 5 miles S.W. off Ailsa Craig. 30 fathoms; clean. Station IX. 4—5 miles W.N.W. from Ailsa Craig and extends 4 miles in a S.W. direction. Depth 20—28 fathoms; mud and sand. Station X. 4 miles long, lies in a N.E. direction, 7 miles from the Cantyre coast, between it and Ailsa Craig. 23 fathoms; sand and sea-weeds. Station XI. passes round Ayr Bay for 4 miles, from 2—4 miles off the shore. Depth 12—20 fathoms; sandy, with rocks and hard ground. Station XII. 4 miles long, extending E. midway between Arran and the Heads of Ayr. 33—40 fathoms; mud[1].

The area of the Clyde, thus closed by the action of the Fishery Board and under the impression that an increase in the number and size of the fishes would result, as well as for experimental purposes, differs in certain features from the Forth and St Andrews Bay, and in others approaches the Moray Frith.

The first period of the "Garland's" work in the Clyde occurred in February and March (1888), and resulted in the capture of 1,367 fishes, of which 705 were saleable and 662 unsaleable. The average for the 13 hauls was 105, or 54 saleable and 51 unsaleable. The small number of hauls (2) in February in proportion to the 11 in March did not reduce the average so much as it would otherwise have done. The difference in the number of food-fishes in this area as contrasted with the Forth was commented on in the Report of the year[2], and it was stated that such was due to the fact that the Forth was closed, whereas in the Clyde trawling was actively prosecuted. But February and March were not the most productive months with which to institute the comparison, since the average for the former was only 27 and for the latter 119. Little weight therefore need be put on this supposition as to the cause of the scarcity, especially as all the hauls occurred in the colder months.

The dab occupies the first place with an average of 23 per haul, a number about a third that of the Moray Frith and St Andrews Bay, but only one less than the average the same

[1] *6th Annual Report, S. F. B.*, 1888, Part III. p. 32.

[2] *6th Annual Report*, Part III. p. 33.

year in the Forth. The dab in the Forth, therefore, had not been benefited by the closure if we entertain the opinion that this fish in the Clyde had suffered from over-fishing. The gurnard is next with 20 per haul, or 7 more than the average in the Forth, while it is 4 less than in St Andrews Bay, and 8 less than in the Moray Frith. The fish next in order is the witch or pole-dab, which has an average of 13 per haul, a number far greater than in the Moray Frith. In the Forth only one or two occur in a haul, while in St Andrews Bay it is absent. This form therefore is, in quantity, characteristic of the area, just as the succeeding one, the hake, is. In the Clyde the average for the hake is 9 per haul, whereas only 1 or 1·5 occur per haul in the Moray Frith, while it is rare in the Forth, at most 1 in 13 hauls, and it is altogether absent from the returns in St Andrews Bay. The lemon-dab was less frequent than in the Forth, only 7 per haul having been obtained, whereas 17 were procured in the Forth the same year, where they were, indeed, 2 above the average of the first quinquennial period, and much more numerous than in St Andrews Bay. The next in order is the whiting with an average of 6 per haul, the smallest of the series, since that for St Andrews Bay was 10, the Moray Frith 22, and the Frith of Forth 27. Plaice come next with an average of 5 per haul, whereas in the Moray Frith it was 14, in the Forth 40, and St Andrews Bay 106. Long-rough dabs were more numerous than in St Andrews Bay, considerably fewer than in the Forth and the Moray Frith. Grey skate were more frequent than in any of the other areas, the next being the Moray Frith. Amongst other fishes sail-flukes occurred more numerously, on the whole, than in the other areas.

The next year in which work was carried on in the Clyde was 1890, and the whole of the hauls (12) were made in July. 1201 fishes, 854, or an average of 71, of which were saleable, and 347, or an average of 29, were unsaleable, the total average being 100, or five less than in 1888. The most abundant form was the witch, the average being 19 per haul, or 6 more than in 1888. The long-rough dab follows with 17 per haul—12 more than in 1888; then the dab with 13, or 11 less than in 1888; the gurnard with an average of 10, or 10 less than in 1888; the whiting with an average of 8, an increase of two per haul; the hake with an average of 6, or three less than in 1888, and the lemon-dab was also diminished. Plaice[1] had considerably diminished, but thornbacks were nearly 3 per haul. The chief feature was the proportionally large number of "witches," and the larger proportion of saleable fishes in contrast with unsaleable. The occurrence of the John Dory and the red gurnard is also worthy of note.

Trawling was not resumed by the "Garland" in the Clyde till 1895, when 12 hauls were made in the month of November, a month which, in the meagre data, stands fifth on the list of 8. The total number captured was 1361, or an average of 113, or 13 over that obtained in July, a fact which shows that the area approaches the condition in the Moray Frith, and diverges considerably from St Andrews Bay and other open seaboards of the east.

The witch had increased its average to 25 per haul, 6 more than in 1890. Gurnards were more than double the number in the latter year, or 24 per haul. Long-rough dabs had also increased. Dabs were 1 more per haul than in 1890, but hake had diminished by 4 per haul and plaice by 13 in a total of 41. Lemon-dabs were fewer, and so were cod and haddock, both of which occurred in small numbers. Grey skate had disappeared from the returns, while thornbacks had increased to fully 4 per haul. A total of 9 soles were captured, and 1 John Dory, 1 smelt and two nurse-hounds showed the peculiarities of the region.

In 1896, 24 hauls of the trawl (in April 12, in October 11, and in November 1) gave a total of 2902 fishes, or 120 per haul, an increase of 7 on the previous year. Of this number 1990 or 82 per haul were saleable, and 912 or 38 per haul unsaleable. An increase of 10 per haul brought the average of the witch to 35, the highest yet obtained. The gurnard followed with 19 per haul, a decrease on the previous year. The average of the long-rough dab had diminished by

[1] An elaborate research on the races of the plaice of Western Europe is now being carried on by Mr H. M. Kyle, M.A., B.Sc., St Andrews.

5 and the plaice had also diminished, but the dab remained the same. Lemon-dabs had increased to 7, and hake to 6 per haul. Cod and haddock were about the same, while the whiting had diminished. Thornbacks were slightly reduced, and soles more distinctly so. Amongst characteristic unsaleable fishes were the John Dory, wrasse, nurse-hound, and black-mouthed dog-fish.

In 1897 the number of hauls was increased to 36, of which 12 occurred in April, 12 in May, and 12 in September. The total number of fishes captured was 4926, or 136 per haul, an increase of 16 on the previous year, and 31 on 1888, a period of 9 years. The gain was not great, but the advocates of the closure are entitled to it, the more so as a comparison of the twelve hauls in April in each year gives the following result:—1896, 114; 1897, 160. An increase of 46 per haul thus took place in the year 1897 in this month. The witches had increased to 47 per haul, a steady rise having taken place from 1888; indeed, their numbers were trebled. The witch, or pole-dab, by some will be held as a typical fish for testing the effects of the closure, since it does not readily take a hook, and as a proof they will cite its steady rise from 13 per haul in 1888 to 47 in 1897. But it is unsafe to trust to an individual form without careful scrutiny of the collateral circumstances. Moreover, the number 13 with which the witch commenced in 1888 could not but be a low number, since the work was carried on only in February and March, two months in which the averages are comparatively small. The two following years had 12 hauls respectively in July and in November, two months with a moderate average, that is, higher than the average of the first year (1888). In 1896 the work was chiefly in April and October, only 1 haul being in November. The average could not fail to be higher than in any previous year—taking into consideration the more productive months dealt with. In the same way, in 1897, the addition of 12 hauls in May to those in April and September still further increased the average, especially of the most abundant fish. There was little in the whole scheme of work to make a safe comparison possible; but, such as it is, it has been shown to be in a line with the results obtained on the east coast.

The pole-dab, again, is not more important than the dab in thus substantiating the benefits of the closure in the Clyde, and yet the dab commenced with 22 per haul in 1888, and concluded with but 12. The plaice began similarly with 5 per haul, and ended with a little over 2. The lemon-dab began with 7 and ended with 5. The whiting began with 6 and ended with 2. A diminution occurred likewise in the cod, sail-fluke, flounder and grey skate. On the other hand, the long-rough dab commenced with 5 per haul and concluded with 17, but this number also occurred in the second period, viz. in 1890; the gurnard began with 20 and concluded with 21, and the hake with 9 and 9. The average captures of the sole during the five years were 11, 5, 9, 8 and 25 per haul, the latter being more than double the number in 1888, but the fact that three low averages intervened shows that an element of uncertainty exists, and that in all probability such an increase is due to the influence of the months of May and September, which are not represented in the previous work of the "Garland" in the Clyde. The haddock showed a slight increase, beginning with 2 and ending with a little over 3. With the exception of the pole-dab and the long-rough dab, it cannot be said that any increase occurred in the flat fishes. The vicissitudes in the returns of the majority of the fishes arise from the usual irregularities, and a consideration of the whole gives little support to the cumulative theory, but rather favours the opinion that food-fishes in general in the area of the Clyde were practically unaffected by the closure.

A similar conclusion is apparent by an examination of the returns for each station. For three of these, without regard to the greatly increased number of hauls, the last entry was lower than the first.

The essential differences between the work of the "Garland" and that of a commercial trawler is clearly brought out by contrasting the results in the protected area of the Clyde with that on the east coast in 1884 in the unprotected waters. Ninety-seven hauls of the "Garland's" trawl in the Clyde area produced a total of 11,757 fishes, of which 8248 were saleable, and 3509 unsaleable, the average in the first being 121, in the second 85, and the third 36. Ninety-three hauls of a commercial trawler on the east coast in 1884 produced no less than 81,854 fishes, of which 69,880 were saleable, and 11,974 unsaleable (with the exception of 361, all being immature), the average in the first being 879, more than 7 times as many as in the Clyde, in the second 751, and in the third 128.

During 1896 and 1897 comparative trials were made of trawling by night and trawling by day on three stations in the Clyde, viz. on VII., VIII. and IX. In 1896 the total of the day-hauls in April on these stations was 320, whereas the total of the night-hauls was 550. In November the 3 hauls on the same stations gave 404 by day and 543 by night. The total was then 724 fishes for the day-hauls and 1093 for the night-hauls. In 1897, however, similar hauls in April on the same stations gave 622 fishes by day and only 485 at night. In May the result was 571 by day and 443 at night; in September 537 by day and 548 at night. For the season the totals were by day 1730 and by night 1476, a different result from the previous year. The totals for the two seasons are for the day 2454, for the night 2569, a slight preponderance in favour of the latter. The experience of fishermen is in favour of the night, and certainly the work on board the beam-trawling ships corroborates this view. The essential point seems to be the presence of darkness, and there may have been circumstances in connection with the ship ("Garland") or the locality, which tended to make the distinctions less pronounced.

CHAPTER VII.

Summary and Conclusions.

The uniformity of the results of the observations and experiments in St Andrews Bay, the Forth, the Moray Frith and the Clyde is worthy of note, for in none can it be said that a substantial increment, or a great diminution has occurred; in none have the fishes increased generally in size, or varied in kinds from what they have always been. Changes may have happened since the days of our forefathers, but they do not seem to be of a kind to be influenced in any marked manner by the restrictive measures hitherto adopted.

There is no diminution in the total quantity landed, but an increase. Nor will it much alter the case to state that this increase has been obtained by a larger fleet of ships and boats, since these, individually, may capture a proportionally smaller number of fishes. In England and Wales a similar condition holds, a progressive increase from 1895 being apparent. Thus in 1896 the total quantity was 7,550,678 cwts. of the value of £5,166,780; while in 1897 the total was 7,946,108 cwts. of the value of £5,568,978. Moreover, haddocks seem to be the fishes of which the largest quantity was taken, viz. 2,548,913 cwts.

In St Andrews Bay, the condition of the fish-fauna immediately after the cessation of trawling showed that the statements as to impoverishment could not be sustained. Not only great abundance of flat fishes, but, in their season, even large round fishes, such as the cod and conger, were captured. The distribution of the food-fishes, indeed, within the bay, remains very much what it was previously, and certainly has not been improved to any noteworthy extent by the closure. With regard to the allegation that frequent trawling on a line (and it would be interesting to know where this line can be so accurately kept) rapidly exhausts the fishes, it is only necessary to point to the statement on p. 99, where it is shown that the fishing-boats of St Andrews from the first introduction of the method to the last day of it found that their richest trawling-line was always the same, viz., Scoonichill in a line with the steeples.

The Frith of Forth showed no increase either in numbers or size in its fish-fauna, but remained at the end of the experiments very much what it was at the beginning. Protection has not increased the food-fishes, nor has it added much to the contentment of the liners, who, it may be, have not fully appreciated their advantages. It is here and there asserted that it is useless to shoot lines in the Forth, *e.g.* in the neighbourhood of the towns of the east, as all the haddocks have disappeared. Now, the "Garland's" work proves that, within easy reach of the coast towns fair captures of haddocks can be made, and which, by energy and perseverance, might be considerably augmented. Moreover, while such is the case, it is also a fact that during the slack period of haddock-fishing a man and two boys may occasionally, in a single night, secure £9 in herring-fishing. Politicians, animated by a desire to benefit the liners by upholding the closure to the three-mile limit, and by extending it if possible to the 13-mile limit, might pause in their efforts if they were aware that it is a moot-point with some liners whether it would not have been better to have confined the trawlers to the three-mile limit, and have given them (the liners) the open sea. No great weight, perhaps, need be attached to such statements, but they show the uncertainty and unrest which appear to be chronic in the department, and which do not seem to have been improved by the methods sometimes adopted. No beneficial effect has ensued in regard to oyster, clam or mussel-bed in the area; indeed, the case of the oyster, which appears to shed spat but rarely in the Forth, or at least which produces little or no viable spat, is still shrouded in mystery. The solution of the difficulties connected with this subject would be a profitable and a worthy field for the energies of the Fishery Board—more productive than a search for benefits from the closure, or a costly and unsatisfactory chase after marauders within the boundaries.

A careful review of the subject and of the fish-fauna in the Moray Frith indicates the absence of substantial data (1) for instituting the closure, and (2) for its continuance. If the results of the experiments in St Andrews Bay and in the Forth had been such as to warrant the closure of a single square mile of the Moray Frith, then the bold step of shutting trawlers out of 2000 square miles in this area would at least have merited the respect due to resolute and masterful methods. So responsible a body as the Fishery Board for Scotland will probably produce its reasons for such a step. So far as they have been disclosed, they are by no means convincing; indeed, a careful search of the entire work of the "Garland," of collateral fishery evidence, and of the history of the Scotch fisheries, fails to afford an explanation. The Fishery Board had the control of a large annual grant for the solution of such questions, and its manipulation of the funds was more or less untrammelled. The results of its experiments must have been either conclusive or inconclusive, unless it is to be supposed that sentiment and not fact pervaded its deliberations. It may be that the laudable view of keeping the peace between the two more pronounced classes of sea-fishermen formed the guiding spirit of the Board's action, but this is a feature at any rate which was not included in the remit by Parliament, as the result of the Royal Commission under Lord Dalhousie. The country has not decided that the closure is to be applied to the whole of the 3-mile limit, or to the 13-mile limit, to keep the peace between the various classes of fishermen. All that has been done is to provide the Board with the means for proving or disproving the allegations of 1883 and 1884, and of satisfying itself and the country as to facts before proceeding to take further steps. Yet what do we find? Only vague statements as the grounds for important measures. Thus, two thousand square miles of the sea were closed, and kept closed, besides schemes for further closures, because—"The conclusions point to the closure of the offshore areas during the spawning time, but the size and precise position of the areas that should be closed in relation to any given part of the territorial waters have not yet been well determined; nor indeed the situations and extent of the principal breeding-grounds[1]." A perusal of the foregoing chapters, and a close scrutiny of the tables at the end of this work, it may be, will enable those who give earnest attention to the subject to reach very different conclusions.

It has also been recently stated[2] that since St Andrews Bay and the Forth are "devoid of spawning-grounds" it is proposed in future to conduct the experiments in the Frith of Clyde and the Moray Frith, where important spawning-grounds exist, and where the results of the closure will therefore be probably much more marked than in the former areas. It is further asserted[3] that "before the investigations of the "Garland" were begun, it was believed that the food-fishes propagated principally in inshore waters, and that an area like the Frith of Forth was in consequence to a large extent self-supporting. But it has been proved that the important food-fishes do not spawn on the east coast within the territorial limits, and that the territorial zone is directly dependent on the spawning grounds situated in the outer seas, whence the floating eggs and fry are borne by the currents." Reference, however, to the Trawling Report of 1884[4], shows that this was clearly brought before the Royal Commission then, so that instead of being new it only corroborates what had been formerly pointed out. Now, what are the results of the ten years' work in the Moray Frith? They simply show that this area is very much as it was in 1884, and that the closure would not appear to affect it more than the open fishing, a conclusion in line with the results of a scrutiny of the Board's work, and independent observations elsewhere. The idea of controlling the "spawning-grounds" of marine fishes with pelagic eggs

[1] 14*th Annual Report*, 1896, Part III. p. 12.
[2] 14*th Annual Report, S. F. B.*, 1896, p. 18.
[3] *Ibid.* p. 12. [4] p. 379, etc.

is chimerical, and cannot be seriously entertained in these days. All the naval resources of Britain would scarcely be adequate to patrol the vast "spawning areas" round our shores, just as the few gunboats and cruisers are insufficient in an area so limited as the Moray Frith. Besides, there are limits in regard to expenditure, and to the patience of the public, though there may be no defined boundary to a "spawning area."

A much more effective method of protecting the fishing-grounds, if such were necessary, would be to limit the number of steam-trawlers, sailing-trawlers and other fishing boats. But while such might be possible in our own country, it would be extremely difficult, if not impossible, to control other nations in this respect.

The result of the closure of the great area included within three miles beyond the Mull of Cantyre, and a similar distance beyond Corsewall Point on the west coast of Wigtownshire, gives no other result than that taught by the three foregoing experiments. No accumulation of food-fishes, no general increase in their size and no alteration of the ways of nature have been effected by the interference of man in this respect.

Thus the great labour and very considerable expenditure during at least ten years prove that the balance of nature in the neighbouring seas is steadily maintained, and that so far as facts go there need be no anxiety concerning the continuance of every species of food-fish.

It is true that in some quarters attempts to minimise the advice of science are made on the plea that science is not practical, but whether this refers to the experience gained from the sale of fishes and shell-fishes on land, or minute acquaintance with their capture at sea, is not clearly stated. The first alternative need not concern us, and the last will give its believers little satisfaction. No class dealing with fishes has probably a more intimate, or more extensive acquaintance with what is, and has been landed by fishermen on our shores and on other shores, with the statistics bearing on this subject, and the reliance to be placed on them, than the scientific investigator in the department. He is familiar with the manipulations of liner, trawler, net-fishing of every kind, and other methods of capture, while, it may be, he is in a position to see daily what is landed in a large or in a small fishing-centre from both the near and the distant grounds. A close observer of nature, he weaves no theories, and is not incautious in deduction. The welfare of the fisheries as a whole is his aim, and the influences which act on those engaged financially in them or have political or other connections with them, are unknown to him.

He considers not only the fishes which haunt the various areas, as made known by the different methods of fishing, but is likewise familiar with the whole contents of the water, from the innumerable floating eggs and swarms of young food-fishes to microscopic plants, all of which as a rule are unknown to those who frequent an area simply to capture as many saleable fishes as possible. He also is cognisant of the history of the subject, and that complaints very similar in nature to those of to-day were made hundreds of years ago, without being followed by the disasters to the food-fishes so frequently reiterated.

Besides the foregoing "practical" qualifications, the scientific observer comes to his task with a precise knowledge of the life-histories of the important food-fishes. He knows, for instance, that the young cod as a rule seeks the inshore waters when about an inch in length, remains near the tidal margin for some time, and as it gets larger and older seeks the offshore grounds[1]. That the haddock, on the contrary, in its early stages is a deep-water fish, and only comes to the inshore waters when 5 or 6 inches in length. He is familiar with the curious cycle in the life-history of the plaice and the turbot, fishes which, as a rule, shed their pelagic eggs in deep water offshore, but the eggs and tiny young are borne shorewards, and are by-and-by to be found at the tidal margin, the former early in the season and in vast numbers, the latter later, larger and in fewer numbers. Further, as these fishes increase in size they return to the deeper waters, thus keeping up as it were a double migration, shorewards in early life, to the offshore when

[1] An interesting and minute study of the growth and early migrations of the cod and whiting has just been completed by Dr A. T. Masterman at St Andrews.

approaching maturity. Having studied as far as possible the life-histories of the food-fishes, he has likewise observed their rate of growth, their habits and so-called migrations, their food, and, last, but not least, their environment in the ocean.

It has still to be shown that such knowledge, which is nothing if not accurate, is a disqualification in dealing with the complex problems of the fisheries. Yet how few officials or public bodies, how few individuals fully realise the prolonged study necessary to grasp all the complex problems of the fisheries. It is, however, not always easy for the public to distinguish between the real and the counterfeit in such a case, and class-interests eagerly seize on the remarks of those who have little more than the rudiments of the subject.

Trawling, the subject of these experiments, appears to have begun systematically at Brixham and Plymouth last century, yet, after more than a hundred years' work, though the area for trawling was by no means extensive, no proof of exhaustion is forthcoming. Long previous to this date, however, viz. in 1376—1377, in the reign of Edward III., a petition was presented to Parliament praying that a new fishing instrument like a large oyster-drag and called a Wondrychroum be interdicted. It was stated that "It destroyed the spawn and brood of fish, and the spat of oysters and mussels and other fish, by which large fish are accustomed to live and be supported. So many small fishes were taken by it that the fishermen did not know what to do with them but feed and fatten their pigs with them to the great damage of the whole commons of the kingdom[1]." Fishes were probably as cheap at this period as formerly. Thus at Billingsgate in the time of Edward I. turbot were sold at 6*d.* each, eels at 8*d.* per cwt., pickled herrings at 1*d.* the score, and oysters at 2*d.* per gallon.

Again, in 1491, during the reign of Henry VII., many of the fishermen of the Norfolk and Suffolk coasts were convicted and fined £10 (=about £100 to-day) for fishing with small-meshed nets and unlawful engines for taking small fishes. So in the reign of James I., and subsequently, fishermen near Rye were "taken" for fishing with unlawful nets,—proving that

[1] Mr Jex, *Fisheries Conference*, Part I. p. 318.

fishermen were determined to deal with the abuses caused by small-meshed nets, especially those used by trawlers. Both foreign and English fishermen were punished for using such nets to the deterioration of the sea-fisheries by capturing the young. The nets were seized, confiscated and burnt. The mesh of the trawl-net was regulated and fixed at various standards in the time of Charles II. Regulations indeed were carried out from the time of James II. to the accession of Queen Anne. Mr Jex (from whom the above is quoted[1]) says abuses were not prominent until the Government ceased to regulate the mesh of the trawl-net.

The need for caution in dealing with the question of the fisheries, indeed, must be apparent to all who have watched the various phases during the last generation. The prominent cry before the Royal Commission on Trawling under Lord Dalhousie was that the decline of the Fisheries was alarming, and that the approach of ruin was near. No reliable statistics other than those of cured fishes were to be had, and thus it was difficult to form an opinion of the exact condition. This alleged serious diminution was coincident with the appearance of steam-trawling in Scotland. It was not previously urged that the sailing-trawlers, in open sandy bays, had ruined the fishing there, and a doubt remains as to whether dissatisfaction so general would have been expressed if the steam-vessels had not entered vigorously into the same pursuit and on the same ground. At any rate, a tendency to use the new seine-trawl has lately appeared in various areas where formerly beam-trawls were employed by the liners.

Many will remember that the drift-net fishermen of the west coast placed the blame of destroying the herring-fishing on the seine-net fishermen. Accordingly in 1860 a close time, on the west coast, during the first five months of the year was enacted for the herrings. Consequently herrings could not be caught for bait in cod-fishing, and starvation and discontent were prevalent, while the herrings were not a whit improved. A Commission, on which were Lord Playfair and Prof. Huxley, removed the restriction with great benefit to the people and to

[1] *Fisheries Conferences*, Part I. p. 318.

the fisheries. The same feature is indicated by M. Georges Roché in regard to the net fishermen of the shores of France, those with drift-nets accusing those with fixed nets as the source of the diminution of the fishes. This author[1] puts other aspects of the case thus: "Si l'on consulte les archives de l'administration de la marine, on voit que de tout temps des plaintes se sont fait entendre à ce sujet. Aux siècles passés, déjà, on prétendait que la mer du Nord était dépeuplée, et l'on accusait certains procédés de travail d'être les causes de ce dépeuplement. Les pétitions adressées au roi, il y a deux siècles, par des pêcheurs de la Méditerranée étaient conçues dans les termes presque identiques à ceux des pétitions actuelles. Tous les règlements qui ont été élaborés pour parer au dépeuplement des mers depuis 1850 sont précédés de considérants tels, que l'on semblait croire, aux époques où ces règlements ont été faits, à une stérilisation presque complète du milieu marin."

The fear which seems to dominate a section of the public is that the sea-fisheries of our offshore and inshore waters may be ruined by over-fishing, as in fresh waters both fluviatile and lacustrine. They do not deem it in consonance with common sense that large ships fitted with the most effective apparatus, and increasing by hundreds annually, should energetically, and in addition to the liners, search those waters without seriously affecting the supply. But they do not keep clearly in view the fact that as soon as an area yields less than what is remunerative either to liner or trawler he changes his ground, and the fishes in addition are scattered. The impoverishment of an area is thus more or less self-curative. It is not within man's power to reduce any area of the sea to utter barrenness, since the constant interchange between it and the vast field of waters is constantly going on. Accordingly, as soon as a certain reduction is made, immigrants by-and-by restore the losses either gradually or by great seasonal shoals. Besides, in connection with trawling, only certain areas are chosen, since sharp rocks or very rough ground are unsuitable. Yet in these sanctuaries the liners can work. The trawl, moreover, captures fishes on or

[1] *La Culture des Mers*, 1898, p. 68.

within a few feet of the ground[1], though the newer kinds may be used in midwater.

It is true that by the closure the lines of the fisherman are thus protected from disturbance and injury by the trawls, and he is left the sole master of the situation, to put in and take out when he chooses. He can spread fleets of lines entirely over bays, and leave them unattended. On the same principles areas may be closed against steam-communication, and traffic relegated to the sailing-boat. In the one case it may happen that lines left all night may not be visited for several days during a storm, whilst hundreds of fishes hang on the hooks all this time or perish from wounds or under attacks of cuttle-fishes, whelks, star-fishes, dog-fishes and other forms. The public, besides, do not obtain so large a supply, especially in rough weather. In the other case, a monopoly of the traffic is inaugurated, with its exorbitant charges and imperfect service.

The complication caused by the anti-trawling agitation has been and is considerable, but is less due to those immediately interested than to interference from the outside and to political circumstances, the former especially being marked wherever limited information exists. It is seldom difficult to rouse class-prejudices especially on a subject so difficult to master as the fisheries, and the network of associations along the seaboard of the country, and generally in communication with each other, is not unfavourable for the rapid extension of such movements.

The protection of the area within the three-mile limit has entailed great responsibility on the Government without in the least altering any of nature's arrangements or adding to the stock of fishes within the areas. Just as in the case of the destruction-of-the-spawn-theory, the demonstration of its fallacy was followed by the destruction-of-the-food-theory, and the destruction-of-the-immature-fish-agitation, by the injury to the fishes themselves and by the devastation of the sea generally; so the closure of certain definite areas for scientific experimental purposes was soon followed by demands for larger areas being closed, then for the whole three-mile limit round the

[1] See Otter-trawl on front book-cover.

coasts, and now for the 10 or 13-mile limit, or perhaps even more. So little can be done by man in the way of altering nature's laws in the sea that it might be more prudent to leave the adjustment of the supply and the demand in her hands.

Some bewail the destruction by trawlers of the rich grounds of the North Sea which they say have been "fished out,"—necessitating voyages to the Bay of Biscay and the west coast of Ireland, and learnedly assert that all fishes have decreased in size since they have not been allowed time to reach maturity. They have found that screw-propellers must kill myriads of young and destroy the floating eggs, though unfortunately the facts are kept in abeyance as being perhaps unnecessary where sensational effect is important.

The situation may be summed up as follows:—the apparent success and general adoption of another method of fishing from that pursued by the liners evoked strenuous opposition. Besides, as time advanced the fishes in many areas seemed to be fewer and more wary, so that the liners had difficulty in making sufficiently large "catches." So many competitors, indeed, are in the field that it is now necessary to proceed further from shore so as to make the pursuit profitable. Instead, however, of placing the onus of the scarcity on any method of fishing, such as trawling, it would be more reasonable to lay the blame on general over-fishing. Certain it is that, with the exception of the herring and the cat-fish, all the important food-fishes have floating or pelagic eggs, and are scatheless from trawls. Nor will the introduction of the paddles and screws of steamers alter the case. Notwithstanding one or two shreds and patches of science, trawled fishes hold their own in every market in our country, and on an average command a higher price than line-caught fishes, though nothing can exceed the condition of the latter. The half-informed outcry about the destruction of the food of the fishes on the bottom has been silenced. The last stronghold, viz. the grave and permanent deterioration of our inshore and offshore fisheries, has been dealt with in the foregoing pages, and, by a practical and unbiassed examination of the areas themselves, and of the statistics of fishes caught in them, the condition is shown to be one that inspires confidence and not alarm. In coming to this conclusion not only the fishes captured for commerce, but the floating eggs in their season, and the minute young fishes, of which no ordinary fisherman has any knowledge, have been duly considered.

Just as in the case of the food-fishes, the antiquity of the coral fishery[1], the comparative freedom in regard to working, and the exhaustion of certain banks, have led to suggestions for restriction. Thus it has been proposed to make a close season during the development of the eggs, but as this period stretches from April to September and even later, the very season when the fishery can best be carried on, the suggestion is impracticable. The proposal of Prof. Lacaze Duthiers to divide the Algerian waters into five areas, each of which should be fished in succession, is noteworthy, though by no means easy to carry out. Such proposals in regard to the red coral of commerce have certainly this to be said in their favour, viz. that the corals are fixed throughout life with the exception of the brief pelagic larval existence. If any form, therefore, would be benefited by the closure of areas, it would be this species, which holds so prominent a place amongst the products of the sea captured by man for gain. Yet, it may be asked, where has this been efficiently and systematically carried out with a corresponding and notable result to the coral fishery?

Suppose, for the sake of argument, that a new kind of fisherman appeared, who, by means of an ingenious instrument, captured the lob-worm in greater numbers and more quickly in the same area, so that it took much harder work for the older fishermen to secure a supply; indeed one of the new fishermen obtained as many in a given time as four of the old. Legislation might, in the same manner as in the case of the trawl, be asked to step in and close the tidal margin against the new fishermen, who would require to pursue their calling by apparatus working under water. Is it to be supposed that, even in this sedentary and comparatively local species, such methods of fishing—the old and the new—would extirpate the lob-worm? It might be more difficult to secure a supply, but those acquainted with the economy of the species, its development, and distribution, would

[1] See Apparatus for Coral-fishing in the Mediterranean on the rear book-cover.

have little fear on this account, however loudly its doom might be pronounced.

In areas constantly harassed there can be no doubt that the fishes become more wary and thus more difficult of capture. A tiny larval or post-larval cod will avoid the forceps intended to capture it, and it is very probable that the adults will move from ground that is constantly fished. The same holds with all the food-fishes. Hence when fishermen find that their captures —on what was rich ground—seriously diminish, such may be due to increased wariness as well as to decimated ranks[1]. Marine fishes differ in habits from fresh-water forms, such as the trout in streams. Instead of from day to day frequenting nearly the same spot with snouts directed to the current, or hiding under the same stones or ledges, the majority of the oceanic round fishes roam hither and thither in vast swarms, consorting with those of a similar size and bound to no locality, for their food is universally distributed. Even the more sedentary flat fishes in their life-cycle often show many changes of habitat, and additional information has lately been given by Dr C. G. J. Petersen from the Limfjord[1] into which the young plaice yearly migrate from the North Sea, and are fished up without a chance of returning in adult life to the open ocean and of reproducing their species.

The larger forms of the food-fishes for the most part are thinned for the time in areas of the open sea closely fished, but the fishes are not extirpated, and if circumstances are favourable they, as a rule, repeople the depleted waters. There are differences however in regard to species. Some, such as the cod, roam after prey, coming inshore in numbers during the winter months and towards the spawning season. Bays have been pronounced as "fished out," yet a single boat, with suitable bait, will at the season mentioned capture 80 fine cod, and others in smaller number. Adult halibut frequent deep water, and appear to become much scarcer when eagerly fished. The smaller forms again are generally found in shallow sandy

[1] *Vide* "On the Memory of Fishes," *Journal of Mental Science*, April, 1898.

[2] "The yearly immigration of Plaice into the Limfjord from the German Sea." *Fisk. Beret.* 1895—96.

bays. Their life-history is not yet fully worked out, but the large forms of 6 ft. would appear to be of considerable age. Hence they are kept in check wherever constant attacks are made. The race, however, continues and is likely to continue.

In some cases it would seem that when the large plaice are fished out of inshore areas, as near Girdleness, they do not return, that is to say, no large plaice have been caught there since. But whether the ground has been carefully examined is not stated. It may be frequented by smaller forms. The large haddocks of Dublin Bay form another instance. They have not returned. But in neither case is the ruin of the fisheries threatened, though these instances should be remembered in dealing with the subject.

In regard to the question, What has been done, in the opportunity afforded by closure, to increase the valuable forms, such as soles and turbot in Scottish waters? two steps have been taken by the Fishery Board, viz. the artificial hatching of plaice, turbot and other forms at Dunbar, and the transference of 500 or 600 soles from English to Scottish waters. The former, that is, the artificial hatching of sea-fishes, is still in the experimental stage, and may have been on too small a scale for efficiency. What the author wrote in 1894 still holds:—

"It has still, of course, to be proved that the artificial hatching of marine fishes, even on a large scale, will be beneficial to the fisheries generally; yet the importance of the issue demands an exhaustive trial. Almost everywhere during the last decade or two complaints have been made as to the decrease of important marine food-fishes. Especially have the large halibut, turbot, brill, soles and other flat fishes become rare. In Britain this alleged diminution has been connected—for the last decade at least—with the extension of beam-trawling in our waters. Consequently, the Legislature has closed the inshore area all round Scottish shores, and even considerably beyond that limit in certain places, as in the Moray Frith and the Frith of Clyde. Many thousand square miles are thus placed solely at the disposal of the liner[1]."

All that can be added is that support should be given to the

[1] *Sc. Progress*, II. p. 252, Dec. 1894.

experiment till the issue is clear, and it is to be hoped that the experiments at Nigg Bay will be definite. The transference of such valuable fishes as the soles might well be extended with benefit to the Scottish fishermen, and without injury to the English waters[1].

Before leaving this part of the subject, it may be remarked that we lately have heard much of the advantages to science of a costly expedition proposed to be sent to the Antarctic regions. The increase of knowledge, even when not directly bearing on the industries of man, always meets with a sympathetic response from our countrymen. But there are so many unsolved problems confronting us in connection with the fisheries at this moment that one is tempted to ask if those in power would not in the first instance do what has most immediate bearing on the welfare of the many thousands depending on the marine fishes for a livelihood. Why should we not be in a position to say, in this nineteenth century, that a fish, say the haddock, extends in great numbers from either hemisphere into the Atlantic, and, if so, whether the pigmy belt of the three-mile or even the thirteen-mile limit can have any more influence on this form than on the ever-abundant herring?

The trawling experiments of the "Garland" have in the foregoing pages been linked on to those carried out for the Royal Commission in 1884, so that continuous observations have been made for fifteen years, and, in certain areas, for a longer period. The aim in all this has been to arrive at reliable results as to the effects of the closure on the captures in the protected areas, on the size of the fishes within the limits, on trawling, and on the welfare of the fisheries generally.

Some are in doubt concerning the sufficiency of time (ten or fifteen years) given to the experiments, asserting that 25 or even 50 years would be necessary, but such arguments rest on no solid basis and are occasionally used *ad*

[1] It was intended in 1894 to transfer a much larger number of soles, but an agitation amongst the fishermen of Yorkshire appeared, and the "Garland" was not permitted to remove more examples. The idea that such would injure the fishing off the Yorkshire coast is untenable. Besides, none of those connected with the experiment would have suggested any measure calculated to be productive of permanent or even temporary prejudice to the local fishermen.

The old (wooden) St Andrews Marine Laboratory, the pleasant home of marine zoologists for twelve years, 1884–1896.

To face p. 232]

captandum. If it were openly stated that such advice was given solely to permit the liners to have the freedom of the inshore waters without molestation, then it could be understood. Whatever may be the opinion on this point it is clear that a period of ten years affords certain important data, especially in conjunction with previous observations, carefully made and free from any disturbing element. Ample time has been given for the growth of the various fishes, from the egg to the adult state.

The conclusion here drawn (and the accompanying tables will be useful in arriving at an opinion) differs from that in the Fishery Board Reports. More especially does this difference affect the statement that "it appears to be fairly well proved that there has been a diminution of the more important flat fishes in the closed waters, instead of an increase, as was anticipated, and that this may probably be traced to the influence of beam-trawling in the open waters where the fishes spawn." The remarks formerly made will show that a more minute analysis of the statistics leads to a different conclusion and one more in harmony with experience.

It is admitted indeed in the Blue-book that no very marked change has taken place in the abundance of the food-fishes in the areas since the prohibition of trawling, yet the plaice and lemon-dab are brought forward to support the view that a "falling off" has occurred. It is the last earthwork in the defence, and seeing that the conditions of the two periods are dissimilar, it lacks that reliability which is indispensable in matters so important. It has already been dealt with on pp. 125—128, and it need only be pointed out here that, by an oversight, it is stated[1] that the information concerning the spawning-grounds of the plaice and lemon-dab was due to the work of the "Garland," but long before the experiments had commenced the spawning-places of both had been determined, viz. during the work for the Trawling Commission in 1884.

The efficacy of a 3 or a 13-mile limit in promoting the increase of the food-fishes is open to doubt. In the first place the food-fishes follow no hard and fast line of distribution so as

[1] e.g. 10*th Annual Report S. F. B.*, p. 12, and subsequently.

to be amenable to such restrictions. A species which extends from the shores of Norway to those of America is probably beyond the influence of such measures. Even were most examples, say of the haddock, removed, for the sake of argument, from the area of the 13-mile limit round the British shores, we have yet to consider the pelagic eggs given off by many of the specimens in the spring months, before and during capture, other eggs swept in by currents from the more distant waters, and the hordes of tiny young which are beyond the reach of any mode of fishing. What permanent influence can a three or a thirteen-mile limit round the small area of the British Islands have on the almost boundless ocean around, an ocean teeming with food from plants to fishes, and harbouring multitudes of the very species we are considering?

Slowly, therefore, the conclusion has been reached that the closure of regions of the open sea in a country like Britain presents few advantages worthy of the constant strain and irritation of class against class, or of the considerable annual expenditure. The capture of great numbers of small fishes by either trawlers or liners is a misfortune for the country, but the closure is powerless to prevent it. Yet so far as history, and so far as observations at the present time go, there is no ground for alarm in regard to the permanence of the food-fishes. Those who have watched the swarms of young soles and other flat fishes on the shrimping-grounds of the Thames, that have been harassed almost daily for hundreds of years, will appreciate the remarks on the permanence of our fish-supply. No test, indeed, could be more decisive, and none more completely answers the half-informed outcry about the disappearance of the haddock from the Scottish waters. Nor is the author alone in this view.

On June 18, 1883, at the Inaugural Meeting of the International Fisheries Exhibition of that year Prof. Huxley, who as a member of several Royal Commissions and Inspector of Salmon Fisheries had large experience, after dealing with the river-fisheries, summed up his address thus:—" And now arises the question, Does the same reasoning apply to the sea-fisheries? Are there any sea-fisheries which are exhaustible, and if so, are the circumstances of the case such that they can be efficiently protected? I believe that it may be affirmed with confidence that, in relation to our present mode of fishing, a number of the most important sea-fisheries, such as the cod fishery, the herring fishery, and the mackerel fishery, are inexhaustible. And I base this conviction on two grounds, first, that the multitude of these fishes is so inconceivably great that the number we catch is relatively insignificant; and, secondly, that the magnitude of the destructive agencies at work upon them is so prodigious, that the destruction effected by the fisherman cannot sensibly increase the death-rate"..."I believe, then, that the cod fishery, the herring fishery, the pilchard fishery, the mackerel fishery, and probably all the great sea-fisheries, are inexhaustible; that is to say that nothing we do seriously affects the number of the fish. And any attempt to regulate these fisheries seems consequently, from the nature of the case, to be useless."

Dr Brown Goode[1] again, whose knowledge of the fisheries of the United States was extensive, states that "the conclusions gained by Prof. Baird, the head of the American Fisheries Staff, tally exactly with those of Prof. Huxley, that the number of any one kind of oceanic fish killed by man is perfectly insignificant when compared with the destruction effected by their natural enemies." He observes of fishes deserting or returning to particular coasts, "their movements are no more to be anticipated than those of the atmosphere; and, in many instances, with no intelligible cause, some of the most abundant species, the blue fish, the chub mackerel, the little tunny, the scuppang, and the bonito, have absented themselves for considerable periods of years." The late Mr Spencer Walpole, whose experience in fishery legislation and of fishermen was great, also agreed with Prof. Huxley. He pointed out however, as a caution, "that though the sea-fisheries are inexhaustible, failure, for which it may not be easy to account or possible to explain, may occur in particular years or during long series of years. Examples of

[1] *Fisheries of the United States*, p. 64.

failures of this character have already been given in relating the history of the Norwegian herring fishery."

Mr Walpole further observes, "It seems, then, reasonable to conclude that oceanic fish, or fish which do not come habitually or necessarily into narrow estuaries or rivers, are incapable of exhaustion by any methods which man has yet invented, or seems likely to invent for their capture. It appears, moreover, that temporary failures of particular fisheries must not be accepted as indications of exhaustion."

He makes sure he cannot be misunderstood thus—"man may rest satisfied that, so far as the open ocean is concerned, the fish which he destroys, if he abstained from destroying them, would perish in other ways[1]."

The Commissioners of 1878 observe, "All[2] sea fish during the earlier stages of their development draw in either to estuaries or to the shallow waters which fringe the shore. But, speaking generally, there is no reason to suppose that the operations of man are making any sensible impression on the number of the fry even in these places, since there is no evidence that the stock of fish in the sea generally is decreasing[3]."

Mr Walpole puts the matter clearly thus, when alluding to Brown Goode's notion that shoals of sea-fish spawning near the coasts may be decimated:—"Even assuming it were possible, I doubt whether any harm would result. No one would think a farmer improvident who brought one-tenth of his herd annually to market. A fish reaches maturity much more rapidly than an ox, and is some thousands of times more productive than a cow. Why then should it be improvident for a fisherman to do what no one would think a farmer improvident for doing? In short, though I doubt the possibility of decimating a shoal of fish, I should regard such a course, if it were practicable, as about the best use the fisherman could make of it[4]."

[1] *Op. cit.* (*Official Report Fish. Exhib.*) p. 138. He excepts seals and whales, crustaceans (crabs and lobsters), oysters, anadromous fishes spawning near the shore, and salmon. *Fisheries Exhib. Lit.* IV. p. 137.

[2] Much has been learned since. The haddock is a marked exception.

[3] *Op. cit.*, p. 152.

[4] *Op. cit.*, p. 153.

He, further, observes, "while I am opposed on the one hand to the imposition of unnecessary restrictions on fishermen, so I am opposed on the other to all patronage simply as such, because I believe the best part of the British fishermen is the independence which they enjoy; and God forbid that the independence which they have won by their own efforts should be taken away from them by the patronage of other people[1]." The same authority pointed out that the prohibition of trawling in the loughs and bays of Ireland had not resulted in an increase of the Irish fisheries[2]. Sir Thomas Brady, lately one of the Inspectors of Irish Fisheries, is of the same opinion.

In connection with Mr Walpole's views, it may be mentioned that not long ago the Government gave grants to the poorer fishermen to aid them in purchasing and improving their boats, but it was found that this did not increase their enterprise and self-reliance, and the experiment has not been continued. On the other hand, no encouragement was needed by the English, Scotch and Manx boats, which drew rich harvests from the waters of their less energetic western brethren, and are ever ready to do so on the same or more distant grounds. No one familiar with the fishermen of our country can have other than the warmest sympathies with them in their daring and uncertain calling, and in their efforts to make the best of a method of fishing which occasionally has difficulty in meeting the requirements of the times, especially against so many competitors. In most pursuits competition is keen, and constant effort is necessary to maintain a good position. There is no need, however, to despair of line-fishing, for there, as elsewhere, skill, steadiness and perseverance will enable its followers to hold their ground.

Mr G. Shaw-Lefevre and Sir J. Caird, members, along with Prof. Huxley, of the Royal Commission of 1866, agreed generally with Prof. Huxley's opinions. That Commission came to the conclusion that there was no ground whatever for the allegation of the falling off of any important fishery on any part of our

[1] *Fisheries Exhib. Lit.* vol. I. p. 130.

[2] *Ibid.* p. 112.

coasts. The same opinion prevailed at that time (1868) in France[1].

Mr Shaw-Lefevre states that from the Scotch returns the average catch of herrings for the 5 years previous to 1863 was 680,000 barrels, and for the 8 years ending 1881, 1,050,000 barrels, an increase of about 50 per cent. During the same period cod and ling had increased about 50 per cent. The number of fishermen had increased by only 15 per cent., the number of boats by the same proportion, while the tonnage had increased by 20 per cent. In 1863 there were 580 trawlers from Hull, Grimsby, Yarmouth, and Ramsgate, in 1881 there were 1500, and in addition the vessels were nearly double the tonnage[2].

He tersely gives his opinion thus:—"I think that it may be taken as an established and incontrovertible fact that there has been no falling off in our fisheries, but a very great increase, an increase in most cases greater in proportion than in the number of men and boats employed, and giving an adequate return to the very largely increased capital devoted to them[3]."

Similar views were held by the late distinguished Prof. J. P. van Beneden of Louvain, who was a member of the Commission appointed by the Belgian Government for investigating the subject of the fisheries. He was of opinion that the products of the sea were inexhaustible, and that no restriction should be placed on the capture of marine fishes. His equally distinguished son, Prof. Edward van Beneden of Liège, Vice-President of the permanent Commission on pisciculture (mariculture), more recently opposed the proposition to restrict the fisheries—broached by the Commission—on the ground that it had not been proved that a general diminution had taken place.

Mr Holdsworth's article on Sea-fisheries in the *Encyclopædia Britannica* breathes the same spirit.

[1] *Internat. Fish. Exhib.* 1883. Conferences, vol. I. p. 88.

[2] *Op. cit.*, pp. 89 and 90.

[3] *Fisheries Exhib. Lit.* IV., Conferences, p. 91, 1884.

M. Georges Roché, the Inspector-in-Chief of the French marine fisheries, on the other hand, accepts the conclusions[1] of the Blue-book in regard to the ten years' work of the Fishery Board as if accurate and final, attributing, for instance, the great increase of 1887 to the accumulation of fishes before the trawlers had time to affect the offshore. He, however, draws attention to the illusion of supposing that the closure of the three-mile limit will ensure the protection of the fisheries.

But it may be said that the increase of the large fishing-vessels[2], both line and trawl, quite alters the situation; that man's skill and the extent of the apparatus now in use must seriously affect the numbers of the food-fishes; and that the condition of to-day is wholly different from that of any previous period in the history of the fisheries.

Similar calamitous forebodings, however, have sporadically, or it may be endemically, appeared in the history of the British fisheries from time immemorial, and have been followed by no disastrous diminution, or the extinction of the most eagerly pursued species. The continuance of the sea-fishes, under every circumstance, has not been generally appreciated, and has not fostered that faith in the marvellous ways of Nature (Divine Providence) in the sea, or built up that confidence for the future which might have been expected as intelligence advanced. The mode of viewing the question has indeed, in some instances, approached that of him who dreaded that the oxygen would by-and-by be diminished in atmospheric air by the apparently boundless increase of man and animals.

A calm survey of the situation, however, shows that the cry concerning the annual diminution of our fish-supply has been dispelled by the institution of statistics; that the alleged destruction of spawn has no basis in fact; that the destruction

[1] *La Culture des Mers*, p. 83, *et seq*.

[2] For instance, 90 were added to those at Grimsby and Hull alone in 1897, making a total of 538 at these two ports. Moreover, there is now a tendency to convert steam liners into trawlers.

of immature fishes is common to all classes of fishermen, and nowhere is proved to have resulted in the ruin of any sea-fishery; that because the first five years of the decade (1886—1895) had a higher average than the second in the Fishery Board's experiments, it therefore followed that diminution of the fishes had occurred, and called for further closures beyond the three-mile limit to remedy it, is shown to rest on insecure data; that the closure of the three-mile limit has failed to increase the number or the size of the food-fishes, is ineffective in regard to the supply of the public, and is a continual source of friction and expense, while falling short of the expectations of those who clamoured for it; that the evidence given before the Trawling Commission of "trawling out" certain grounds in three years with a small vessel carrying a small trawl, the working period being about three times a week for three months in autumn, is at variance with experience; that the statements to the effect that fishes captured by the trawl are inferior as articles of food to the general public cannot be maintained either by science or by a knowledge of the markets; that the "Garland's" work shows the comparatively small destruction of immature fishes of value, even though she often trawled where no commercial ships would; that the perusal of masses of fishery-statistics shows the constant series of changes that take place on every area, yet the fisheries are not destroyed; that such a fishery as that for sparlings in the estuary of the Tay has from time immemorial been very much as it is; that though salmon and sea-trout abound in the sea man derives little knowledge of their presence by either trawl or hook, and yet many of both must come in their way.

Lastly, the results of the foregoing investigations do not seem to offer a basis for the fitting-out of fast cruisers with search-lights to pursue the trawling vessels of our countrymen in their own waters. They may be of importance, however, in preserving order between the various classes of fishermen, and in carrying out properly organised scientific experiments and observations in connection with the food-fishes, which, it is clear, have not received any appreciable benefit from the closure.

The returns from the various centres all over the country have for the most part steadily increased since 1884[1], and though it is true that large quantities are captured on the Great Fisher Bank, Iceland, and other regions at a distance from British waters proper, yet this is due to the more remunerative nature of the work, and not to the dearth of fishes in the seas at home. Such increase has not been fostered by the closure, but, in the case of both liner and trawler, is due to enterprise which was independent.

Nature has been able by her unaided resources to ward off extinction in a species so eagerly desired by man as the red or precious coral for one of the greatest incentives of the race, viz. pecuniary gain, and yet so circumscribed in distribution, and so slenderly supplied with means of dispersion in comparison with many marine animals. Moreover, all this has occurred in a sea, which from its limited boundaries is specially excluded from the consideration of the question treated of in this work, and which has been swept by hundreds of boats' crews annually by day and by night. If the continuance of the red coral, therefore, has been assured, even if it be granted (for authorities deem the diminished price rather than scarcity of coral at the root of the present depression) that its numbers have been lessened, what difficulty is there in regard to the permanent abundance of the chief food-fishes of the open seaboard of our country, set as it is in the midst of an almost boundless ocean, with all the marvellous powers of increase, a thousand-fold greater than the coral, so characteristic of them on the one

[1] The long-continued abundance of the food-fishes of Britain may be estimated by the following note on a year's supply to Billingsgate about 1850 (*Chambers' Miscellany*, "The Commerce of the Thames"):—The *Morning Chronicle* "3 or 4 years ago" gave a list of the fishes brought in a year to Billingsgate. Live cod 400,000; barrelled cod 750,000; salt cod 1,600,000; salmon 200,000; fresh haddocks 2,500,000; smoked haddocks 19,500,000; soles 97,000,000; mackerel 23,000,000; fresh and red herrings and bloaters 200,000,000; eels 10,000,000; whitings 20,000,000; plaice 36,000,000; brills and mullets 1,200,000; turbot 800,000; oysters 500,000,000; lobsters and crabs 2,000,000; shrimps and prawns 500,000,000. Besides, even now, so large an amount as 37 tons may be destroyed as unfit for food in a month.

hand, and all the varied and gigantic resources of nature at command on the other? Science as well as experience answers that there is none.

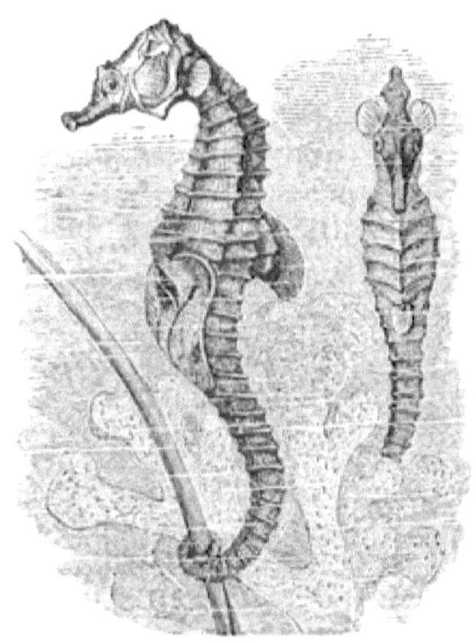

Sea-horses (*Hippocampi*). The male with his brood-pouch is on the left.

INDEX.

CAMBRIDGE: PRINTED BY J. AND C. F. CLAY, AT THE UNIVERSITY PRESS.

An oblique view of the Gatty Marine Laboratory from the land-face (west).

To face p. 242]

TABLE I.

"GARLAND'S" TRAWLING IN ST ANDREWS BAY.

1886—1890, 25 periods. 1891—1895, 29 periods.

	Jan.	Feb.	March	April	May	June	July	Aug.	Sept.	Oct.	Nov.	Dec.	Colder	Warmer
1886	—	—	—	—	—	5	2	—	5	—	5	—	5	12
1887	—	—	—	—	5	5	—	5	—	5	—	—	—	20
1888	—	—	—	—	—	—	5	5	5	5	—	5	5	20
1889	4	—	5	5	—	5	—	5	—	5	5	—	19	15
1890	—	—	5 5†	—	5	—	—	5	—	5	—	5	15	15
Monthly No. of Hauls 1886—1890	4	—	15	5	10	15	7	20	10	20	10	10	44	82
Average No. per Haul 1886—1890	16	—	94	132	270	365	292	615	303	200	146	47	126	
1891	—	5	—	—	—	—	5 5	5	—	5	5	3	13	20
1892	—	5	5	5	5	5	—	—	—	5	5	5	25	15
1893	5	5	5	—	—	5	5	—	—	—	5	—	20	10
1894	—	5	—	5	—	—	—	—	5	—	—	5	15	5
1895	—	—	5	5	—	5	—	—	—	5	—	—	10	10
Monthly Hauls in second period	5	20	15	15	5	15	15	5	5	15	15	13	83	60
Average No. per Haul on total 1891—1895	13	35	68	119	214	403	186	124	483	257	221	34	143	
Total No. of Hauls in each month	9	20	30	20	15	30	22	25	15	35	25	23	269	
	79				142						48		127	142
Average (for years 1886—1895) of *all* fishes* per Haul	14	35	81	122	251	384	220	517	363	224	191	40		

* Saleable, immature saleable, and unsaleable.
† Two examinations—each with 5 Hauls.

TABLE II.

TABLE SHEWING NUMBER OF SALEABLE FISHES

per month from 1886 to 1895.

"GARLAND'S" TRAWLING IN ST ANDREWS BAY.

	No. of Hauls	Jan.	Feb.	March	April	May	June	July	Aug.	Sept.	Oct.	Nov.	Dec.	Totals
		*5	5	15	5	10	15	17	25	10	25	14	13	
1886	17	—	—	—	—	—	977	48	—	496	—	271	—	1792
1887	21	—	—	—	—	738	1149	—	471	—	291	—	—	2649
1888	25	—	—	—	—	—	—	1834	1756	1517	986	—	126	6219
1889	35	63	—	463	630	—	1247	—	1914	—	805	904	—	6026
1890	30	—	—	903	—	1222	—	—	3837	—	332	—	127	6421
1891	33	—	173	—	—	—	—	1099	524	—	357	1408	111	3672
1892	44	—	87	86	313	676	914	—	—	—	630	196	111	3013
1893	32	35	53	423	—	—	872	916	—	—	—	550	—	2849
1894	24	—	324	—	889	—	—	—	—	1972	—	—	92	3277
1895	23	—	—	234	587	—	1870	—	—	—	2184	—	—	4875
		98	637	2109	2419	2636	7029	3897	8502	3985	5585	3329	567	

* No. of Hauls.

TABLE III.

TABLE SHEWING NUMBER OF UNSALEABLE FISHES

per month from 1886 to 1895.

"GARLAND'S" TRAWLING IN ST ANDREWS BAY.

	No. of Hauls	Jan.	Feb.	March	April	May	June	July	Aug.	Sept.	Oct.	Nov.	Dec.	Totals
		*5	5	15	5	10	15	17	25	10	25	14	13	
1886	17	—	—	—	—	—	313	48	—	688	—	137	—	1186
1887	21	—	—	—	—	673	1495	—	1991	—	1366	—	—	5525
1888	25	—	—	—	—	—	—	115	90	337	25	—	27	594
1889	35	2	—	21	34	—	285	—	399	—	107	156	—	1004
1890	30	—	—	29	—	72	—	—	1858	—	95	—	193	2247
1891	33	—	49	—	—	—	—	{256 91	98	—	149	921	97	1661
1892	40	—	131	110	149	395	988	—	—	—	254	318	131	2476
1893	32	31	8	56	—	—	777	620	—	—	—	56	—	1548
1894	24	—	142	—	232	—	—	—	—	533	—	—	54	961
1895	23	—	—	124	62	—	939	—	—	—	950	—	—	2075
		33	330	340	477	1140	4797	1130	4436	1558	2946	1588	502	

* No. of Hauls.

TABLE IV.

"GARLAND'S" TRAWLING IN ST ANDREWS BAY.

STATIONS I. TO V. — STATION VI.

No. of Hauls	Year	Average per Haul (Saleable and Unsaleable Fishes) at each Station					Totals			Average per Haul at all Stations			Station VI. Average per Haul				Station VI. Totals		
		I	II	III	IV	V	Saleable Fishes	Unsaleable Fishes	Grand Total	Saleable Fishes	Unsaleable Fishes	Saleable and Unsaleable Fishes	No. of Hauls	Saleable Fishes	Unsaleable Fishes	Saleable and Unsaleable Fishes	Saleable Fishes	Unsaleable Fishes	Grand Total
17	1886	163	73	213	264	168	1792	1186	2978	105	69	175	—	—	—	—	—	—	—
20	1887	479	409	276	571	307	2649	5525	8174	132	276	408	—	—	—	—	—	—	—
25	1888	178	243	390	329	221	6218	595	6813	248	23	272	—	—	—	—	—	—	—
34	1889	245	224	125	264	178	6010	1020	7030	176	30	206	—	—	—	—	—	—	—
30	1890	254	221	183	721	64	6395	2273	8668	213	75	288	—	—	—	—	—	—	—
126	1886—1890	254	238	227	436	179	23064	10599	33663	183	84	267	—	—	—	—	—	—	—
33	1891	131	118	136	142	292	3668	1665	5333	111	50	161	—	—	—	—	—	—	—
40	1892	125	77	152	159	105	2680	2277	4957	67	56	123	4	82	50	133	331	201	532
30	1893	102	111	115	265	93	2713	1405	4118	90	46	137	2	66	73	139	133	146	279
20	1894	159	125	145	287	156	2763	737	3500	138	36	175	4	127	57	184	509	229	738
20	1895	307	249	249	507	254	4469	1802	6271	223	90	313	3	135	91	226	406	273	679
143	1891—1895	151	123	153	242	162	16293	7886	24179	113	55	169	13	106	65	171	1379	849	2228*
269	1886—1895	199	178	187	332	171	39357	18485	57842	146	68	215	—	—	—	—	—	—	—

* 1892—1895 for Station VI.

DREWS BAY. 1886—1895.

Grey Skate	Angler	Sandy Ray	Sail-Fluke	Thornback	Starry Ray	Conger	Flapper Skate	Herring	Catfish	Average per Haul: Plaice	Dab	Haddock	Gurnard
16	—	1	—	—	7	—	1	—	—	70	61	9	24
65	—	—	—	4	6	—	10	—	1	151	130	50	52
2	—	—	—	44	5	—	—	—	—	106	94	34	20
8	14	2	—	42	4	—	—	—	—	121	55	—	18
3	36	—	—	32	15	—	—	2	1	169	73	—	21
94	50	3	—	122	37	—	11	2	2	127	80	16	25
7	5	1	—	13	14	1	—	86	2	48	86	—	13
2	5	1	—	30	17	—	—	176	1	41	31	—	27
5	3	2	—	12	12	—	—	4	1	59	24	27	13
2	16	—	1	10	12	—	—	—	—	57	71	21	8
7	39	—	—	30	5	1	—	—	—	86	148	15	44
23	68	4	1	95	60	2	—	266	4	56	64	12	20
117	118	7	1	217	97	2	11	268	6	89	72	14	22

TABLE V.

GARLAND'S* TRAWLING IN ST ANDREWS BAY. 1886—1895.

STATIONS I. TO V.

No. of Hauls	Year	Totals																						Average per Haul			
		Plaice	Dab	Lemon Dab	Long Rough Dab	Cod	Haddock	Whiting	Gurnard	Flounder	Turbot	Brill	Sole	Grey Skate	Angler	Sandy Ray	Sail-Fluke	Thornback	Starry Ray	Conger	Flapper Skate	Herring	Catfish	Plaice	Dab	Haddock	Gurnard
17	1886	1204	1038	24	65	14	159	43	431	22	4	1	—	16	—	1	—	—	7	—	1	—	—	70	61	9	24
20	1887	3035	2605	105	86	4	1040	33	1048	142	13	—	4	65	—	—	—	4	6	—	22	—	1	151	130	50	52
25	1888	2650	2373	12	65	6	863	238	517	4	8	—	—	2	—	—	—	61	5	—	—	—	—	106	94	34	20
34	1889	4142	1880	10	30	15	32	43	633	138	7	2	1	8	14	2	—	61	4	—	—	—	—	122	55	—	18
50	1890	5070	2200	5	22	45	45	388	658	122	3	—	2	3	36	—	—	52	25	—	—	2	1	102	73	—	21
126	1886—1890	16100	10144	156	266	86	2109	745	3167	428	35	3	7	94	50	5	—	122	37	—	11	2	2	127	80	16	25
35	1891	1598	2865	4	22	20	56	114	649	41	1	—	2	7	5	1	—	13	84	1	—	86	2	48	86	—	13
48	1892	1971	1241	10	76	45	104	336	1098	217	9	8	1	2	3	1	—	30	17	—	—	176	2	41	30	—	27
30	1893	1786	704	12	76	75	834	91	400	66	3	—	—	5	3	2	—	12	12	—	—	4	1	59	24	27	13
20	1894	1145	1436	3	17	17	422	85	179	89	1	1	—	2	16	—	1	20	12	—	—	—	—	57	71	21	8
20	1895	1733	2973	15	12	12	310	38	892	240	1	3	—	7	39	—	—	30	5	1	—	—	—	86	148	15	44
143	1891—1895	7935	9219	44	203	172	1726	684	3021	453	15	12	3	23	68	4	1	95	60	2	—	266	4	56	64	12	20
269	1886—1895	24035	19370	200	471	258	3835	1429	6188	877	48	15	10	117	118	7	1	217	97	2	11	268	6	89	72	14	22

NOVE

889	1891	1
5	5	
—	2	
—	2	
—	2	
—	—	
—	—	
—	3	
—	—	
—	—	
—	—	
—	1	
—	2	
—	1	
—	—	
—	8	
3	5	
—	—	
—	—	
—	—	
—	—	
1	—	
2	1	
—	—	
—	—	
—	—	
—	1	
114	74	
677	170	
4	1	
63	1041	
—	—	
—	2	
13	—	
1	—	
23	59	
3	31	
—	—	
—	—	

GARLA

	JUN			DECEMBER					Total of each Species	SPECIES	Average per Haul of each Species
	87	1889	1892	890	1891	1892	1894	Totals			
	5	5	5	5	3	5	5				269
G	—	—	—	—	—	—	—	—	42	Grey Skate	·15
Th	—	2	—	—	1	—	—	1	56	Thornback	·20
St	—	—	—	1	—	—	2	5	58	Starry Ray	·21
Sa	—	1	—	—	—	1	—	1	4	Sandy Ray	·01
H	—	—	—	1	—	2	—	3	208	Herring	·77
C	—	—	—	2	—	—	—	2	24	Cod	·09
H	92	—	—	—	2	45	1	48	1193	Haddock	4·43
W	—	—	3	176	21	18	1	218	772	Whiting	2·86
W	—	—	—	—	—	—	—	—	19	Whiting-Pout	·07
Li	1	—	—	—	—	—	—	—	1	Ling	·003
Lo	9	—	1	1	—	4	10	16	137	Long-Rough Dab	·51
T	—	—	—	—	—	—	—	—	14	Turbot	·05
Br	1	—	—	—	—	—	—	—	3	Brill	·01
Pl	65	27	386	—	—	1	—	1	4246	Plaice	15·41
D	82	201	356	4	59	11	3	100	9323	Dab	34·65
Le	1	—	—	—	—	1	—	1	112	Lemon-Dab	·41
So	—	—	—	—	—	—	—	—	2	Sole	·007
Fl	—	—	—	—	—	—	—	—	65	Flounder	·24
G	32	49	240	2	—	—	—	2	1947	Grey Gurnard	7·23
C	—	—	—	—	—	—	—	—	1	Catfish	·003
Sa	—	—	—	—	—	2	—	2	2	Sand-Eel	·007
Sp	—	—	—	3	14	—	—	17	40	Sprat	·14
Fl	2	—	—	—	—	—	—	—	10	Flapper-Skate	·03
P	—	—	—	—	—	—	—	—	1	Poor Cod	·003
Sa	95	280	986	190	97	85	17	417 500	18280		
M								917 39			

	JUN				DECEMBER					Total of each Species	SPECIES	Average per Haul of each Species
	87	1889	1892	88	1890	1891	1892	1894	Totals			
	5	5	5	5	5	3	5	5				269
An	—	9	2	—	—	—	—	1	1	118	Angler	·43
Sk	—	1	—	—	3	—	2	—	5	27	Skulpin	·10
A	—	—	—	—	—	—	1	1	2	28	Armed Bullhead	·10
G	—	—	—	—	—	—	—	—	—	1	Goby	·003
L	—	1	—	—	—	—	—	—	—	6	Lumpsucker	·02
Sh	—	1	—	—	—	—	1	—	1	12	Short-spined Cottus	·04
B	—	3	—	—	—	—	—	—	—	10	Butterfish	·03
Fi	—	—	—	—	—	—	—	—	—	2	Five-Bearded Rockling	·007
M	—	—	—	—	—	—	—	—	—	1	Montagu's Sucker	·003
	—	15	2	—	3	—	4	2	9	205		

TABLE VIII.

"GARLAND'S" TRAWLING IN FRITH OF FORTH.

1886—1890, 34 periods. 1891—1895, 53 periods.

No. of Hauls		Jan.	Feb.	March	April	May	June	July	Aug.	Sept.	Oct.	Nov.	Dec.	Colder	Warmer
27	1886	—	—	—	—	—	9	3	—	5	4	6	—	6	21
24	1887	—	—	—	—	—	9	—	9	6	—	—	—	—	24
50	1888	—	—	—	—	—	8	—	8	9	9	7	9	16	34
90	1889	9	9	9	9	9	9	6	12	—	9	9	—	45	45
79	1890	—	9	9	9	9	—	7	6	5	9	7	9	43	36
270	Total No. of Hauls	9	18	18	18	18	35	16	35	25	31	29	18	110	160
	Average No. per Haul on Total	69	81	92	101	139	230	372	325	350	262	144	133	270	
	1886—1890														
106	1891	9	9	7	4	14	9	9	9	7	11	9	9	47	59
99	1892	9	9	9	5	13	9	9	—	9	9	7	11	50	49
96	1893	8	8	2	9	8	10	7	11	8	9	7	9	43	53
71	1894	6	9	9	9	—	9	—	9	9	—	6	5	44	27
75	1895	3	9	5	9	9	—	9	9	9	6	—	7	33	42
447	Average No. per Haul on Total 1891—1895	125	137	117	166	194	237	354	493	354	265	153	127	217	230
	Total No. of Hauls	35	44	32	36	44	37	34	38	42	35	29	41	447	
	Total No. of Hauls	44	62	50	54	62	72	50	73	67	66	58	59	327	390
	Average No. per Haul on Total 1886—1895	114	121	108	144	178	233	360	413	354	264	148	129	717	

Only the nine original stations calculated.

TABLE IX.

"GARLAND'S" TRAWLING IN FRITH OF FORTH.

No. of Hauls	Year	Average per Haul (saleable and unsaleable) at each Station									Totals			Average per Haul		
		I	II	III	IV	V	VI	VII	VIII	IX	Saleable Fishes	Unsaleable Fishes	Totals	Saleable	Unsaleable	Totals
27	1886	211	308	423	204	169	197	196	114	46	4649	1247	5896	172	46	218
24	1887	566	453	513	304	401	154	358	249	342	3257	5837	9094	135	243	378
50	1888	205	309	227	266	157	81	216	195	84	8928	1024	9952	178	20	198
90	1889	173	155	172	272	146	112	119	121	102	11981	1787	13768	133	19	152
79	1890	209	233	314	139	274	115	316	300	180	14953	3317	18270	189	42	231
270	1886—1890	231*	249	286	229	214	123	225	197	126	43768	13212	56980	162	48	211
106	1891	212	196	209	248	164	95	203	99	86	13325	4687	18012	125	44	169
99	1892	202	185	189	159	225	112	214	131	107	9394	7342	16736	93	74	167
96	1893	210	209	282	233	411	154	528	273	261	14130	13271	27401	147	138	285
71	1894	203	220	296	285	613	113	349	212	216	15349	4232	19581	216	59	275
75	1895	286	268	440	324	341	71	385	194	128	16960	3610	20570	226	48	274
447	1891—1895	220*	213	271	243	328	112	330	179	154	69158	33142	102300	154	74	228
717	1886—1895	225†	226	277	238	283	116	290	308	144	112926	46354	159280	157	64	222

* Quinquennial average at each Station.
† Decennial average.

ARL.

TAL

Total Sole	Total White Skate	Witch	Total Pollack	Total Sprat	Total Halibut	Dog-fish	John Dory	Sea-Trout	Zeugo-pterus
(1)	—	—	—	—	—	—	—	—	—
(1)	—	—	—	—	—	—	—	—	—
—	(5)	(57)	(1)	(4)	—	—	—	—	—
—	—	(102)	(11)	(5)	—	—	—	—	—
(1)	—	(54)	—	(15)	(1)	(1)	—	—	—
(3)	(5)	(213)	(12)	(24)	(1)	(1)	—	—	—
(2)	—	(71)	—	(60)	—	(3)	(1)	—	—
—	—	(67)	—	(13)	(1)	—	—	(1)	(1)
(1)	—	(102)	—	(11)	(1)	—	—	—	—
(3)	—	(123)	—	(2)	—	(1)	—	—	—
(1)	—	(98)	(2)	(1)	—	(2)	—	—	—
7	—	461	2	87	2	6	1	1	1
10	5	674	14	111	3	7	1	1	1
—	—	—	—	—	—	—	—	—	—
—	—	—	—	—	—	—	—	—	—
—	—	—	—	—	—	—	—	—	—
—	—	—	—	—	—	—	—	—	—

Table X.

GA

SE

92	1893	1890	189
	10	5	—
1	4	3	—
—	2	—	—
—	3	—	—
7	—	—	—
—	1	—	—
—	—	—	—
—	—	—	—
5	—	—	—
9	3	7	—
5	282	41	—
7	1	10	—
5	34	304	—
4	9	304	—
4	—	6	—
2	53	64	—
4	38	609	—
—	—	—	—
—	—	—	—
—	2	—	—
—	—	—	—
—	—	—	—
—	1	1	—
7	8	5	—
2	25	4	—
4	369	91	—
—	—	1	—
—	—	—	—
1	7	—	—
5	611	111	—
1	60	10	—
3	2	1	—
3	47	120	—
5	22	37	—
—	—	—	—
8	—	—	—
2	—	—	—
—	26	38	—
—	12	13	—
—	—	—	—
5	1	—	—
1	6	—	—
—	—	—	—
—	—	—	—
—	—	—	—
—	—	—	—
—	—	—	—
—	—	—	—
—	—	—	—
—	—	—	—
3	17 129	1780	—

Avera
er Ha
of eac
Speci

717

·12
·16
·17
·0
·2
1·2
20·8
3·6
·0
·0
·0
·0
—
—
·0
12·8
·0
·0
1·7
15·9
1·9
·0
·0
3·3
·0
·0
·1
·0
62·6

Aver
per H
of e
Spec

71

1·

NG.

DECEMBER								SPECIES
1889	1890	1891	1892	1893	1894	1895	Totals	
—	9	9	11	9	5	7	59	
—	1	—	—	—	—	2	3	Grey Skate
—	6	3	1	—	2	1	14	Thornback
—	5	1	2	—	—	5	14	Starry Ray
—	—	—	—	—	—	—	—	Sandy Ray
—	8	—	—	3	—	—	11	Herring
—	44	1	—	81	16	1	143	Cod
—	7	48	440	40	—	1	537	Haddock
—	58	45	33	86	7	—	237	Whiting
—	1	—	—	—	—	—	1	Whiting-Pout (Bib)
—	—	—	—	—	—	—	—	Coalfish
—	—	1	—	—	—	—	1	Hake
—	1	—	—	1	—	—	2	Ling
—	—	—	—	—	—	—	—	Halibut
—	—	—	—	—	—	—	—	Sail-Fluke
—	—	—	—	—	—	—	—	Craig-Fluke (Witch)
—	99	128	152	72	175	59	735	Long-Rough Dab
—	—	—	—	—	—	—	—	Turbot
—	—	—	—	—	—	—	—	Brill
—	—	—	1	—	—	—	1	Plaice
—	114	94	11	12	13	25	277	Dab
—	3	7	7	1	1	—	19	Lemon-Dab
—	—	—	1	—	—	—	1	Sole
—	—	—	—	—	—	—	—	Flounder
—	4	4	1	—	—	—	9	Grey Gurnard
—	—	—	—	—	—	—	—	Bream
—	—	—	—	—	—	—	—	Catfish
—	—	—	—	—	—	—	—	Flapper-Skate
—	—	—	—	—	—	—	—	Sand-Eel
—	2	—	—	—	—	—	2	Sprat
—	2	4	2	—	—	—	8	Brassie?
—	355	336	651	296	214	94	2015	

UN

DECEMBER

"ROYAL SAXON'S" TRAWLING IN MORAY FRITH. 1884.

OFF COAST OF CAITHNESS.

(Smith Bank.)

No. of Hauls 3.

TABLE XIII.

SALEABLE FISHES.

SPECIES	APRIL			Totals
	I	II	III	
	8th	8th	9th	
Grey Skate	1	—	—	1
Starry Ray	1	—	—	1
Cod (large)	2	21	15	38
Do.	12	31	21	64
Haddock (large)	354	961	497	1812
Whiting	5	7	9	21
Ling	4	—	—	4
Halibut	—	1	—	1
Witch	3	—	—	3
Turbot	—	—	1	1
Brill	—	1	1	2
Plaice	15	100	78	193
Dab	7	21	14	42
Lemon-Dab	29	97	26	152
Gurnard	37	160	152	349
Catfish	11	12	4	27
Totals	481	1412	818	2711

Average per Haul 903.

TABLE XIV.

UNSALEABLE FISHES.

SPECIES	APRIL			Totals
	I	II	III	
	8th	8th	9th	
Angler	3	3	2	8
Totals	3	3	2	8

Y FRITH. 1898.

TABLE XVI.

MMATURE (UNSALEABLE) FISHES.

SPECIES							
	APRIL						
	I	II	III	IV	V	VI	Totals
	7th	7th	8th	8th	8th	8th	
....	16	—	—	—	5	9	30
....	36	24	6	10	19	50	145
....	—	—	—	—	—	1	1
....	9	7	—	14	4	8	42
....	—	—	—	2	—	—	2
....	1	—	—	—	—	—	1
....	18	—	—	—	—	—	18
ab	—	5	—	7	9	19	40
....	5	—	—	—	—	—	5
....	11	2	9	33	20	4	79
....	1	1	1	—	5	—	8
	97	39	16	66	62	91	371
....	14	—	—	—	—	—	14
....	111	39	16	66	62	91	385

ix young Cod (small) were liberated alive.

SPECIES

Grey Skate...
Cod ....
Do. ....
Do.(Codling)
Haddock......
Do. ....
Do. ....
Whiting ....
Saithe ....
Ling ....
Witch ....
Turbot.........
Do. ....
Plaice ....
Sail-Fluke ...
Dab............
Lemon-Dab
Gurnard ....
Catfish.........
Poor Cod

Totals ......

No. of Hauls 6.

TABLE XV.

SALEABLE FISHES.

SPECIES	Description	APRIL						Totals
		I	II	III	IV	V	VI	
		7th	7th	8th	8th	8th	8th	
Grey Skate		—	1	1	—	2	1	5
Cod	Large	24	—	—	—	—	—	24
Do.	Medium	—	8	2	13	9	4	36
Do.(Codling)	Small	—	1	—	4	5	13	23
Haddock	Large	203	184	243	136	249	120	1135
Do.	Medium	311	180	134	102	261	97	1085
Do.	Small	341	232	175	219	484	357	1808
Whiting		12	9	14	31	19	32	117
Saithe		—	—	—	—	2	—	2
Ling		—	—	5	2	2	1	10
Witch		32	7	—	—	8	5	52
Turbot	Large	—	—	1	—	—	—	1
Do.		—	—	—	—	1	1	2
Plaice		1	—	6	6	5	—	18
Sail-Fluke		34	2	1	—	11	9	57
Dab		—	—	—	—	8	—	8
Lemon-Dab		14	1	23	19	17	45	119
Gurnard		24	—	53	110	86	14	287
Catfish		—	—	5	4	2	4	15
Poor Cod		—	2	—	5	—	—	7
Totals		993	727	663	642	1171	719	4914

TABLE XVI.

IMMATURE (UNSALEABLE) FISHES.

SPECIES*	APRIL						Totals
	I	II	III	IV	V	VI	
	7th	7th	8th	8th	8th	8th	
Grey Skate	16	—	—	—	5	9	30
Starry Ray	36	24	6	10	19	50	145
Sandy Ray	—	—	—	—	—	1	1
Haddock	9	7	—	14	4	8	42
Whiting	—	—	—	2	—	—	2
Hake	1	—	—	—	—	—	1
Witch	18	—	—	—	—	—	18
Long-Rough Dab	—	5	—	7	9	19	40
Sail-Fluke	5	—	—	—	—	—	5
Dab	11	2	9	33	20	4	79
Lemon-Dab	1	1	1	—	5	—	8
	97	39	16	66	61	91	371
Angler	14	—	—	—	—	—	14
Totals	111	39	16	66	61	91	385

* Six young Cod (small) were liberated alive.

NOVEMBER

1892	1893	1894	1895	1896	1897	Totals	Totals of each Species	SPECIES
—	—	—	—	6	6	12	108	
—	—	—	—	—	—	—	2	Grey Skate (large)
—	—	—	—	—	—	—	9	Do.
—	—	—	—	—	2	2	65	Thornback (large)
—	—	—	—	2	3	5	66	Do.
—	—	—	—	—	—	—	—	Starry Ray
—	—	—	—	—	—	—	1	Sandy Ray
—	—	—	—	—	1	1	1	Fuller's Ray
—	—	—	—	—	—	—	1	Shagreen-Ray
—	—	—	—	2	—	2	32	Cod (large)
—	—	—	—	46	19	65	234	Do.
—	—	—	—	6	17	23	121	Haddock (large)
—	—	—	—	22	18	40	449	Do.
—	—	—	—	—	—	—	430	Do. (small saleable)
—	—	—	—	—	1	1	6	Whiting (large)
—	—	—	—	—	4	4	181	Do.
—	—	—	—	—	2	2	127	Do. (small saleable)
—	—	—	—	—	—	—	7	Hake
—	—	—	—	1	1	2	3	Ling
—	—	—	—	—	—	—	1	Lythe (Pollack)
—	—	—	—	—	—	—	—	Halibut
—	—	—	—	—	—	—	1	Sail-Fluke
—	—	—	—	6	42	48	388	Witch
—	—	—	—	—	2	2	31	Long-Rough Dab (large)
—	—	—	—	7	20	27	321	Do.
—	—	—	—	—	1	1	6	Turbot
—	—	—	—	10	3	13	34	Brill
—	—	—	—	6	26	32	320	Plaice (large)
—	—	—	—	84	163	247	1678	Do.
—	—	—	—	57	221	278	3315	Do. (small saleable)
—	—	—	—	—	1	1	62	Dab (large)
—	—	—	—	39	182	221	3104	Do.
—	—	—	—	10	15	25	319	Lemon-Dab
—	—	—	—	—	—	—	20	Sole
—	—	—	—	—	—	—	11	Flounder (large)
—	—	—	—	—	—	—	18	Do. (small saleable)
—	—	—	—	1	6	7	251	Gurnard (large)
—	—	—	—	—	7	7	1012	Do.
—	—	—	—	—	—	—	2	Catfish (large)
—	—	—	—	—	—	—	1	Do.
—	—	—	—	299	757	1056	12630	

—	—	—	—	6	Grey Skate
—	1	3	4	23	Thornback
—	—	—	—	1	Starry Ray
—	3	—	3	15	Cod
—	—	—	—	343	Haddock
—	—	1	1	45	Whiting
—	—	—	—	8	Hake
—	—	3	3	11	Witch
—	14	17	31	167	Long-Rough Dab
—	—	—	—	2	Turbot
—	—	63	63	1227	Plaice
—	30	354	384	4837	Dab
—	—	—	—	32	Lemon-Dab
—	—	—	—	6	Sole
—	—	—	—	4	Flounder
—	1	6	7	348	Gurnard
—	—	1	1	2	Poor Cod
—	—	1	1	5	Solenette
—	—	3	3	13	Butterfish
—	—	—	—	5	Pipefish
—	—	—	—	1	Wrasse
—	8	16	24	99	Angler
—	—	6	6	12	Dragonet
—	—	3	3	5	Pogge
—	—	—	—	7	Fatherlasher
—	—	—	—	8	?
—	—	1	1	1	Four-bearded Roc[illegible]

TABLE XIX.

"GARLAND'S" TRAWLING. MORAY FRITH. 1887—1897.

STATIONS I. TO VI.

Year	April	May	June	July	August	Sept.	Oct.	Nov.	Colder Months	Warmer Months	No. of Hauls
1887	—	—	—	—	6	—	—	—	—	6	6
1888	—	6	—	—	—	—	—	—	—	6	6
1889	—	—	6	—	—	—	—	—	—	6	6
1890	—	6	—	6	—	6	—	—	—	18	18
1891	—	—	—	—	—	6	—	—	—	6	6
No. of Hauls 1887—1891	—	12	6	6	6	12	—	—	—	42	42
Average No. per Haul 1887—1891	—	147	105	136	261	199	—	—	—	170	170
1892	—	—	—	—	—	6	—	—	—	6	6
1893	1	5	—	—	—	—	6	—	1	11	12
1894	—	—	—	6	—	—	6	—	—	12	12
1895	—	—	—	5	1	—	—	—	—	6	6
1896	—	—	—	—	6	—	6	6	6	12	18
1897	—	—	6	—	—	—	—	6	6	6	12
No. of Hauls 1892—1897	1	5	6	11	7	6	18	12	13	53	66
Average No. per Haul 1892—1897	35	91	89	172	87	293	321	132	125	208	192
Average No. per Haul 1887—1897	35	131	97	160	167	230	321	132	125	192	183
No. of Hauls 1887—1897	1	17	12	17	13	18	18	12	13	95	108

TABLE XX.

"GARLAND'S" TRAWLING. MORAY FRITH. 1887—1897.

STATIONS I. TO VI.

No. of Hauls	Year	Average per Haul (Saleable and Unsaleable Fishes) at each Station						Totals			Average per Haul at all Stations		
		I	II	III	IV	V	VI	Saleable Fishes	Unsaleable Fishes	Grand Total	Saleable Fishes	Unsaleable Fishes	Saleable and Unsaleable Fishes
6	1887	49	542	50	227	349	350	600	967	1567	100	161	261
6	1888	200	315	29	105	398	111	1108	50	1158	184	8	193
6	1889	90	131	159	124	79	51	546	88	634	91	14	105
18	1890	81	199	35	181	251	60	1433	993	2426	79	55	134
6	1891	45	393	69	245	491	152	875	520	1395	145	85	230
42	Five Years 1887—1891	89	282	58	178	296	120	4562	2618	7180	108	62	170
6	1892	196	275	34	1238	4	16	1043	720	1763	173	120	293
12	1893	197	271	64	436	451	499	2195	1669	3864	182	139	322
12	1894	131	170	109	474	271	353	2139	879	3018	178	73	251
6	1895	91	142	33	94	263	92	380	335	715	63	55	119
18	1896	122	66	61	72	75	120	1201	352	1553	66	19	86
12	1897	141	186	63	80	173	241	1110	660	1770	92	55	147
66	Six Years 1892—1897	144	170	64	320	207	241	8068	4615	12683	122	69	192

7—1897.

No. of Hauls	Year	Angler	Plaice	Dab	Haddock	Gurnard	
			Average per Haul				
6	1887	7	93	61	53	30	August
6	1888	4	14	33	23	39	May
6	1889	4	47	18	10	20	June
18	1890	9	30	72	4	8	May, July, Sept.
6	1891	4	37	141	3	18	September
42	1887—1891 Five years	28	40	68	14	19	
6	1892	3	174	85	9	10	September
12	1893	8	117	131	31	22	April, May, Oct.
12	1894	16	94	113	10	13	July, October
6	1895	7	22	62	8	10	July, August
18	1896	16	30	31	2	5	Aug., Oct., Nov.
12	1897	21	45	61	3	11	June, November
66	1892—1897 Six years	71	73	77	11	12	

TABLE XXI.

"GARLAND'S" TRAWLING. MORAY FRITH. 1887—1897.

STATIONS I. TO VI.

No. of Hauls	Year	Totals																Average per Haul				
		Plaice	Dab	Lemon-Dab	Long Rough Dab	Cod	Haddock	Whiting	Gurnard	Hake	Turbot	Brill	Witch	Flounder	Grey Skate	Thornback	Angler	Plaice	Dab	Haddock	Gurnard	
6	1887	559	370	42	14	5	318	20	182	—	2	2	22	11	11	—	7	93	61	51	30	August
6	1888	88	201	30	132	7	143	135	215	4	—	—	172	1	1	5	4	14	33	23	35	May
6	1889	282	110	16	—	3	61	1	125	—	1	—	—	—	—	3	4	47	18	10	20	June
18	1890	542	1337	29	68	38	75	118	155	—	1	—	33	5	—	9	9	30	74	4	8	May, July, Sept.
6	1891	222	846	27	51	25	18	18	110	8	—	—	49	5	—	6	4	37	141	3	18	September
42	1887—1891 Five years	1693	2864	144	265	78	615	292	807	12	4	2	276	22	12	23	28	40	68	14	19	
6	1892	1049	511	14	4	29	59	5	61	—	2	2	1	4	—	14	3	174	85	9	10	September
12	1893	1415	1577	33	53	29	379	36	272	—	—	2	19	—	—	13	8	117	131	31	22	April, May, Oct.
12	1894	1135	1361	74	42	33	123	4	167	—	—	7	7	2	3	33	16	94	113	10	13	July, October
6	1895	136	375	18	23	5	52	10	69	—	—	—	1	4	—	18	7	22	62	8	10	July, August
18	1896	548	574	26	54	72	70	6	99	3	—	18	23	1	2	39	16	30	31	2	5	Aug., Oct., Nov.
12	1897	549	733	42	78	45	45	8	140	—	2	3	72	—	—	14	21	45	61	3	11	June, November
66	1892—1897 Six years	4828	5131	207	254	203	728	67	804	3	4	32	123	11	5	131	71	73	77	11	12	

SPECIES	Si[ze] In[s]	[OC]TOBER 1896	1897	Totals	NOVEMBER 1893	1894	1895	1896	1897	Totals	Totals of each Species
		10	—	30	—	—	—	4	9	13	87
Grey Skate (large)	16	1	—	2	—	—	—	—	2	2	12
Do.	12	—	—	2	—	—	—	—	—	—	2
Thornback (large)	19	1	—	3	—	—	—	1	2	3	13
Do.	13	1	—	2	—	—	—	1	2	3	10
Starry Ray	11	1	—	11	—	—	—	—	2	2	25
Fuller's Ray	11	—	—	—	—	—	—	—	1	1	1
Sandy Ray	11	—	—	4	—	—	—	1	—	1	7
Cod (large)	23	—	—	7	—	—	—	2	3	5	20
Do.	8	6	—	75	—	—	—	2	53	55	219
Haddock (large)	15	2	—	19	—	—	—	2	13	15	104
Do.	10	512	—	2898	—	—	—	188	105	293	4075
Do. (small saleable)	8	3	—	902	—	—	—	7	5	12	1599
Whiting (large)	14	2	—	15	—	—	—	—	3	3	29
Do.	10	24	—	235	—	—	—	2	14	16	384
Do. (small saleable)	7	1	—	59	—	—	—	—	3	3	102
Hake	12	8	—	13	—	—	—	1	1	2	22
Ling	15	—	—	7	—	—	—	2	—	2	10
Halibut	13	—	—	—	—	—	—	—	—	—	1
Sail-Fluke	7	2	—	9	—	—	—	1	2	3	26
Witch		10	—	85	—	—	—	—	49	49	196
Long-Rough Dab (large)	10	1	—	1	—	—	—	—	3	3	6
Do.	8	69	—	183	—	—	—	11	90	101	433
Turbot	7	1	—	1	—	—	—	1	—	1	3
Brill	7	—	—	1	—	—	—	—	1	1	5
Plaice (large)	19	3	—	24	—	—	—	1	4	5	66
Do.	12	22	—	157	—	—	—	1	33	34	348
Do. (small saleable)	7	1	—	5	—	—	—	—	—	—	11
Dab (large)	12	—	—	2	—	—	—	—	1	1	7
Do.	7	407	—	1592	—	—	—	201	548	749	3158
Lemon-Dab	7	73	—	262	—	—	—	15	64	79	627
Flounder (large)	9	—	—	—	—	—	—	—	—	—	—
Do.	7	—	—	—	—	—	—	—	—	—	—
Gurnard (large)	11	141	—	335	—	—	—	5	31	36	467
Do.	7	257	—	717	—	—	—	10	169	179	1562
Catfish (large)	33	—	—	—	—	—	—	—	—	—	3
Do.	15	—	—	—	—	—	—	—	—	—	5
Conger	33	2	—	2	—	—	—	—	—	—	2
		1641	—	7630	—	—	—	455	1204	1659	13560

TABLE XXII.

"GARLAND'S" TRAWLING. 1893—1897. MORAY FRITH.

STATIONS VII. TO XVI. SALEABLE FISHES.

SPECIES	Size in Inches	MAY				JUNE		JULY						AUGUST						OCTOBER						NOVEMBER						Totals of each Species
		1893	1894	1895	Totals	1897	Totals	1893	1894	1895	1896	1897	Totals	1893	1894	1895	1896	1897	Totals	1893	1894	1895	1896	1897	Totals	1893	1894	1895	1896	1897	Totals	
		10*	—	—	10	22	22	—	10	1	—	—	11	—	—	9	4	—	13	10	10	—	12	—	32	—	—	—	4	9	13	97
Grey Skate (large)	16—	2	—	—	2	1	1	—	3	—	—	—	3	—	—	2	1	—	3	1	—	—	1	—	2	—	—	—	—	2	2	12
Do.	11—15	—	—	—	—	—	—	—	—	—	—	—	—	—	—	—	—	—	—	—	2	—	—	—	2	—	—	—	—	—	—	2
Thornback (large)	19—	2	—	—	2	—	—	—	4	—	—	—	4	—	—	1	—	—	1	1	1	—	1	—	5	—	—	—	1	2	3	13
Do.	13—18	1	—	—	1	—	—	—	4	—	—	—	4	—	—	—	—	—	—	—	1	—	1	—	2	—	—	—	1	2	3	10
Starry Ray	11—	5	—	—	5	6	6	—	—	—	—	—	—	—	—	2	1	—	3	5	5	—	1	—	11	—	—	—	—	2	2	25
Tuller's Ray	11—	—	—	—	—	—	—	—	—	—	—	—	—	—	—	—	—	—	—	—	—	—	—	—	—	—	—	—	—	1	1	1
Sandy Ray	16—	—	—	—	—	—	—	—	1	—	—	—	1	—	—	—	—	—	—	1	3	—	—	—	4	—	—	—	1	—	1	7
Cod (large)	23—	2	—	—	2	2	2	—	3	—	—	—	3	—	—	1	—	—	1	1	6	—	—	—	7	—	—	—	2	1	3	20
Do.	8—22	62	—	—	62	3	3	—	5	—	—	—	5	—	—	8	11	—	19	42	25	—	6	—	73	—	—	—	2	51	53	215
Haddock (large)	15—	58	—	—	58	9	9	—	5	—	—	—	5	—	—	—	—	—	—	14	3	—	2	—	19	—	—	—	4	13	17	104
Do.	10—14	280	—	—	280	145	145	—	143	37	—	—	179	—	—	344	43	—	387	282	1204	—	512	—	1998	—	—	—	188	105	293	4073
Do. (small saleable)	8—9	90	—	—	90	43	43	—	275	32	—	—	307	—	—	43	16	—	59	300	338	—	3	—	902	—	—	—	7	3	12	1300
Whiting (large)	14—	4	—	—	4	1	1	—	4	—	—	—	4	—	—	3	1	—	4	3	8	—	2	—	13	—	—	—	—	3	3	29
Do.	10—13	24	—	—	24	6	6	—	8	18	—	—	26	—	—	64	16	—	79	14	107	—	84	—	235	—	—	—	2	14	16	384
Do. (small saleable)	7—9	3	—	—	3	—	—	—	15	2	—	—	17	—	—	20	—	—	20	11	47	—	1	—	59	—	—	—	—	3	3	102
Hake	18—	—	—	—	—	2	2	—	2	1	—	—	3	—	—	2	1	—	3	2	3	—	8	—	13	—	—	—	2	2	2	22
Ling	15—	—	—	—	—	1	1	—	—	—	—	—	—	—	—	—	—	—	—	3	4	—	—	—	7	—	—	—	2	—	2	10
Halibut	13—	—	—	—	—	—	—	—	—	—	—	—	—	—	—	1	—	—	1	—	—	—	—	—	—	—	—	—	—	—	—	2
Sail Fluke	7—	2	—	—	2	4	4	—	3	—	—	—	3	—	—	—	4	—	4	—	2	—	3	—	9	—	—	—	1	2	3	26
Witch		1	—	—	1	1	1	—	30	4	—	—	34	—	—	13	10	—	23	22	34	—	10	—	65	—	—	—	—	49	49	190
Long Rough Dab (large)	10—	1	—	—	1	1	1	—	—	—	—	—	—	—	—	—	—	—	—	—	—	—	1	—	4	—	—	—	—	1	3	6
Do.	8—9	13	—	—	13	72	72	—	2	—	—	—	2	—	—	36	22	—	58	15	99	—	69	—	183	—	—	—	11	90	101	434
Turbot	5—	—	—	—	—	—	—	—	—	—	—	—	—	—	—	—	1	—	1	—	—	—	2	—	2	—	—	—	1	—	1	3
Brill	7—	1	—	—	1	—	—	—	2	—	—	—	2	—	—	—	—	—	—	—	1	—	—	—	1	—	—	—	—	1	1	5
Plaice (large)	10—	17	—	—	17	6	6	—	9	—	—	—	9	—	—	3	—	—	5	10	11	—	3	—	24	—	—	—	1	4	5	66
Do.	12—18	36	—	—	36	31	31	—	95	1	—	—	96	—	—	38	6	—	44	88	47	—	22	—	157	—	—	—	1	33	34	368
Do. (small saleable)	8—12	3	—	—	3	1	1	—	—	—	—	—	—	—	—	2	1	—	3	3	1	—	1	—	5	—	—	—	—	—	—	12
Dab (large)	12—	2	—	—	2	2	2	—	—	—	—	—	—	—	—	—	—	—	—	3	—	—	—	—	3	—	—	—	—	1	1	7
Do.	8—12	121	—	—	121	266	266	—	125	27	—	—	152	—	—	174	104	—	278	443	852	—	407	—	1702	—	—	—	200	348	708	3158
Lemon Dab	7—	51	—	—	51	79	79	—	69	2	—	—	71	—	—	72	14	—	85	205	86	—	73	—	364	—	—	—	15	64	79	667
Flounder (large)	9—	—	—	—	—	—	—	—	—	—	—	—	—	—	—	—	—	—	—	—	—	—	—	—	—	—	—	—	—	—	—	—
Do.	7—8	—	—	—	—	—	—	—	—	—	—	—	—	—	—	—	—	—	—	—	—	—	—	—	—	—	—	—	—	—	—	—
Gurnard (large)	11—	28	—	—	28	1	1	—	27	1	—	—	29	—	—	17	15	—	32	74	130	—	140	—	333	—	—	—	3	51	36	467
Do.	7—10	165	—	—	165	108	108	—	150	18	—	—	168	—	—	180	43	—	225	265	295	—	157	—	717	—	—	—	10	169	179	1562
Catfish (large)	33—	1	—	—	1	2	2	—	—	—	—	—	—	—	—	—	—	—	—	—	—	—	—	—	—	—	—	—	—	—	—	3
Do.	15—32	1	—	—	1	4	4	—	—	—	—	—	—	—	—	—	—	—	—	—	—	—	—	—	—	—	—	—	—	—	—	5
Conger	33—	—	—	—	—	—	—	—	—	—	—	—	—	—	—	—	—	—	—	—	—	—	2	—	2	—	—	—	—	—	—	4
		879	—	—	879	802	802	—	1133	139	—	—	1272	—	—	1005	313	—	1318	1366	4423	—	1641	—	7630	—	—	—	433	1204	1659	13580

* Number of Hauls.

1893—1897.

HES.

	OCTOBER						NOVEMBER						Totals of each Species
tals	1893	1894	1895	1896	1897	Totals	1893	1894	1895	1896	1897	Totals	
3	10	10	—	10	—	30	—	—	—	4	9	13	87
—	—	2	—	—	—	2	—	—	—	—	—	—	2
—	—	1	—	1	—	2	—	—	—	1	1	2	9
—	1	4	—	1	—	6	—	—	—	1	—	1	11
—	—	—	—	1	—	1	—	—	—	—	—	—	2
1	1	—	—	—	—	1	—	—	—	1	—	1	14
84	393	—	—	3	—	396	—	—	—	—	6	6	974
—	—	—	—	—	—	—	—	—	—	—	3	3	7
2	—	—	—	3	—	3	—	—	—	1	3	4	9
1	—	—	—	—	—	—	—	—	—	—	—	—	1
—	—	1	—	—	—	1	—	—	—	—	—	—	1
58	158	342	—	132	—	632	—	—	—	145	216	361	1907
—	—	—	—	—	—	—	—	—	—	—	—	—	—
17	702	298	—	524	—	1524	—	—	—	264	1315	1579	4719
15	7	7	—	3	—	17	—	—	—	—	7	7	56
79	73	19	—	40	—	132	—	—	—	21	48	69	431
—	8	—	—	10	—	18	—	—	—	1	5	6	29
—	1	—	—	—	—	1	—	—	—	—	—	—	2
—	—	—	—	—	—	—	—	—	—	1	—	1	1
22	14	31	—	11	—	56	—	—	—	12	15	27	148
—	7	—	—	1	—	8	—	—	—	—	3	3	15
1	2	1	—	—	—	3	—	—	—	—	—	—	8
—	—	—	—	—	—	—	—	—	—	—	—	—	1
—	—	—	—	—	—	—	—	—	—	—	—	—	1
80	1367	706	—	730	—	2803	—	—	—	448	1622	2070	8348

TABLE XXIII

"GARLAND'S" TRAWLING. MORAY FRITH. 1893—1897.

STATIONS VII. TO XVI. UNSALEABLE FISHES.

SPECIES	Size in Inches	MAY						JUNE		JULY						AUGUST						OCTOBER						NOVEMBER						Totals of each Species
		1893	1894	1895	1896	1897	Totals	1897	Totals	1893	1894	1895	1896	1897	Totals	1893	1894	1895	1896	1897	Totals	1893	1894	1895	1896	1897	Totals	1893	1894	1895	1896	1897	Totals	
		10*	—	—	—	—	10	10	10	—	10	1	—	—	11	—	—	9	4	—	13	10	10	—	10	—	30	—	—	—	4	9	13	87
Grey Skate	—18	—	—	—	—	—	—	—	—	—	—	—	—	—	—	—	—	—	—	—	—	—	2	—	—	—	2	—	—	—	—	—	—	2
Thornback	—13	1	—	—	—	—	1	1	1	—	1	—	—	—	1	—	—	—	—	—	—	—	1	—	1	—	2	—	—	—	1	1	2	7
Starry Ray	—11	2	—	—	—	—	2	1	1	—	—	—	—	—	—	—	—	—	—	—	—	1	6	—	1	—	8	—	—	—	1	—	1	12
Sandy Ray	—11	—	—	—	—	—	—	1	1	—	—	—	—	—	—	—	—	—	—	—	—	—	—	—	1	—	1	—	—	—	—	—	—	2
Cod	—8	11	—	—	—	—	11	—	—	—	—	—	—	—	—	—	—	1	—	—	1	1	—	—	—	—	1	—	—	—	1	—	1	14
Haddock	—8	489	—	—	—	—	489	3	3	—	4	12	—	—	16	—	—	84	—	—	84	283	—	—	3	—	286	—	—	—	—	6	6	884
Whiting	—7	1	—	—	—	—	1	—	—	—	1	—	—	—	1	—	—	—	—	—	—	—	—	—	—	—	—	—	—	—	—	3	3	7
Hake	—12	—	—	—	—	—	—	—	—	—	—	—	—	—	—	—	—	—	2	—	2	—	—	—	3	—	3	—	—	—	1	3	4	9
Halibut	—13	—	—	—	—	—	—	—	—	—	—	—	—	—	—	—	—	1	—	—	1	—	—	—	—	—	—	—	—	—	—	—	—	1
Sail-Fluke	—7	—	—	—	—	—	—	—	—	—	—	—	—	—	—	—	—	—	—	—	—	—	1	—	—	—	1	—	—	—	—	—	—	1
Long Rough Dab	—8	48	—	—	—	—	48	124	124	—	318	68	—	—	386	—	—	130	128	—	258	158	342	—	130	—	630	—	—	—	145	116	261	1707
Plaice	—7	—	—	—	—	—	—	—	—	—	—	—	—	—	—	—	—	—	—	—	—	—	—	—	—	—	—	—	—	—	—	—	—	—
Dab	—7	306	—	—	—	—	306	312	312	—	484	77	—	—	561	—	—	300	117	—	417	700	298	—	586	—	1584	—	—	—	164	1215	1379	4559
Lemon-Dab	—7	6	—	—	—	—	6	5	5	—	4	2	—	—	6	—	—	14	1	—	15	7	7	—	3	—	17	—	—	—	—	7	7	56
Gurnard	—7	64	—	—	—	—	63	15	15	—	57	—	—	—	57	—	—	48	31	—	79	13	78	—	40	—	131	—	—	—	41	48	89	434
Poor Cod	—	5	—	—	—	—	5	—	—	—	—	—	—	—	—	—	—	—	—	—	—	8	—	—	10	—	18	—	—	—	1	5	6	29
Solenette	—	1	—	—	—	—	1	—	—	—	—	—	—	—	—	—	—	—	—	—	—	1	—	—	—	—	1	—	—	—	—	—	—	2
Butterfish	—	—	—	—	—	—	—	—	—	—	—	—	—	—	—	—	—	—	—	—	—	—	—	—	—	—	—	—	—	—	1	—	1	1
Angler	—	9	—	—	—	—	9	20	20	—	14	—	—	—	14	—	—	20	1	—	20	14	31	—	11	—	56	—	—	—	12	15	27	148
Dragonet	—	1	—	—	—	—	1	1	1	—	3	—	—	—	3	—	—	—	—	—	—	7	—	—	1	—	8	—	—	—	—	3	3	16
Pogge	—	—	—	—	—	—	—	3	3	—	1	—	—	—	3	—	—	—	1	—	1	1	1	—	—	—	3	—	—	—	—	—	—	8
Wolf-fish	—	—	—	—	—	—	—	—	—	—	1	—	—	—	1	—	—	—	—	—	—	—	—	—	—	—	—	—	—	—	—	—	—	1
Three-bearded Rockling	—	—	—	—	—	—	—	—	—	—	1	—	—	—	1	—	—	—	—	—	—	—	—	—	—	—	—	—	—	—	—	—	—	1
		948	—	—	—	—	948	498	498	—	890	159	—	—	1049	—	—	698	282	—	980	1367	708	—	730	—	2805	—	—	—	448	1612	2060	8340

* Number of Hauls.

TABLE XXIV.

"GARLAND'S" TRAWLING. MORAY FRITH. 1893—1897.

STATIONS VII. TO XVI.

Year	May	June	July	Aug.	Oct.	Nov.	Colder Months	Warmer Months	Total No. of Hauls
1893	10	—	—	—	10	—	—	20	20
1894	—	—	10	—	10	—	—	20	20
1895	—	—	1	9	—	—	—	10	10
1896	—	—	—	4	10	4	4	14	18
1897	—	10	—	—	—	9	9	10	19
No. of Hauls 1893—1897	10	10	11	13	30	13	13	74	87
Average No. per Haul 1893—1897	182	130	211	176	347	286	286	245	251

TABLE XXV.

"GARLAND'S" TRAWLING. MORAY FRITH. 1893—1897.

STATIONS VII. TO XVI.

No. of Hauls	Year	Average per Haul (Saleable and Unsaleable Fishes) at each Station										Totals			Average per Haul at all Stations		
		VII	VIII	IX	X	XI	XII	XIII	XIV	XV	XVI	Saleable Fishes	Unsaleable Fishes	Grand Total	Saleable Fishes	Unsaleable Fishes	Saleable and Unsaleable Fishes
20	1893	428	280	216	121	183	321	173	117	238	301	2445	2315	4760	122	115	238
20	1894	479	624	582	283	70	68	214	452	440	361	5556	1596	7152	277	79	357
10	1895	298	614	274	95	104	117	86	86	159	168	1144	857	2001	114	85	200
18	1896	325	257	164	193	221	193	44	77	251	262	2409	1460	3869	133	81	214
19	1897	415	315	300	205	144	160	86	128	273	70	2006	2120	4126	105	111	217
87	1893—1897	392	382	296	189	140	176	134	194	289	260	13560	8348	21908	155	95	251

893—1897.

N… H…	Sandy Ray	Angler	Average per Haul: Plaice	Dab	Lemon-Dab	Long-Rough Dab	Haddock	Whiting	Gurnard	
3	1	23	7	69	8	11	97	3	28	May, October
9	5	45	5	87	8	38	158	13	33	July, October
2	—	20	4	57	8	33	52	10	26	July, August
1	2	25	2	94	5	28	43	2	31	Aug., Oct., Nov.
3	1	35	4	128	8	26	17	1	20	June, November
5	9	148	3·9	90	7	26	77	6	28	
1	—	8	64	14	50	—	604	7	116	April
5	1	14	3	14	21	6	695	19	47	April

TABLE XXVI.

GARLAND'S* TRAWLING. MORAY FRITH. 1893—1897.

STATIONS VII. TO XVI.

No. of Hauls	Year	Totals: Plaice	Dab	Lemon-Dab	Long Rough Dab	Cod	Haddock	Whiting	Gurnard	Hake	Ling	Sail-Fluke	Witch	Turbot	Brill	Flounder	Grey Skate	Thornback	Starry Ray	Sandy Ray	Angler	Average per Haul: Plaice	Dab	Lemon-Dab	Long Rough Dab	Haddock	Whiting	Gurnard	
20	1893	136	1300	167	237	118	1948	63	578	2	3	1	20	—	1	—	[illegible]	5	11	1	63	7	69	8	12	97	3	28	May, October
29	1894	213	1759	106	764	43	3171	109	668	4	4	13	84	—	3	—	6	11	9	3	45	5	83	8	38	139	13	32	July, October
10	1895	46	578	89	338	10	527	106	265	3	—	—	17	—	—	—	1	1	8	—	20	4	57	8	33	52	10	26	July, August
18	1896	55	1707	106	906	22	776	46	365	15	2	7	20	3	—	—	3	6	4	4	65	3	94	5	65	43	2	15	Aug., Oct., Nov.
19	1897	75	2464	115	506	62	309	28	358	6	1	6	33	—	1	—	3	7	8	4	35	4	128	8	26	17	1	90	June, November
63	1893—1897	425	7884	683	2346	253	6752	352	2460	31	10	27	196	3	5	—	16	31	36	9	148	3·9	90	7	26	37	6	28	
†3	1889	193	42	139	—	102	2818	11	345	—	6	—	3	1	8	—	6	—	4	—	5	64	14	50	—	609	7	116	April
16	1898	18	87	177	90	89	4173	819	287	1	10	64	76	1	—	—	33	—	145	1	94	3	14	11	6	695	10	47	April

* On Smith Bank, &c., inside Moray Frith.
† Outside Moray Frith.

DE.

S.

R	OCTOBER	NOVEMBER			Totals of each Species
	1896	1895	1896	Totals	
	11	12	1	13	97
	—	—	—	—	4
	—	—	—	—	11
	2	10	—	10	53
	14	18	1	19	102
	10	7	—	7	24
	1	—	—	—	2
	—	—	—	—	1
	—	—	—	-	8
	—	—	—	—	1
	—	—	—	—	5
	5	6	—	6	34
	2	5	—	5	39
	7	16	6	22	170
	18	—	11	11	74
	—	10	—	10	22
	6	22	2	24	219
	3	—	—	—	102
	28	26	2	28	404
	1	2	1	3	12
	—	—	—	—	8
	—	—	—	—	62
	9	8	—	8	92
	338	309	1	310	3285
	—	2	—	2	74
	32	79	—	79	373
	2	3	—	3	10
	4	2	—	2	20
	2	4	—	4	24
	16	15	—	15	212
	5	9	1	10	41
	1	—	1	1	10
	48	77	27	104	628
	46	32	36	68	509
	4	9	—	9	58
	—	—	—	—	6
	—	—	—	—	25
	127	89	—	89	490
	122	187	6	193	985
	3	—	—	—	32
	—	8	—	8	8
	—	—	—	—	1
	1	2	—	2	8
	857	957	95	1052	8428

TABLE XXVII.

GARLAND'S* TRAWLING. FRITH OF CLYDE.

SALEABLE FISHES. TOTALS FOR FIVE YEARS.

SPECIES	Size in Inches	FEBRUARY	MARCH	APRIL			MAY	JULY	SEPTEMBER	OCTOBER	NOVEMBER			Totals of each Species
		1888	1888	1896	1897	Totals	1897	1899	1895	1895	1899	1896	Totals	
		2*	11	12	12	24	12	12	12	11	12	1	13	97
Grey Skate (large)	16—	—	—	—	1	1	1	1	1	—	—	—	—	4
Do.	12—15	—	9	1	—	1	—	—	1	—	—	—	—	11
Thornback (large)	19—	—	—	3	8	11	8	10	12	2	10	—	10	53
Do.	13—18	—	—	6	15	21	18	12	18	14	18	1	19	102
Sandy Ray	11—	—	—	6	—	6	1	—	—	10	7	—	7	24
Flapper Skate	11—	—	1	—	—	—	—	—	—	1	—	—	—	2
Shagreen Ray	11—	—	—	—	1	1	—	—	—	—	—	—	—	1
Cuckoo Skate	11—	—	—	—	—	—	—	4	4	—	—	—	—	8
Raia maculata	11—	—	—	—	1	1	—	—	—	—	—	—	—	1
Cod (large)	23—	—	—	1	—	1	2	1	1	—	—	—	—	5
Do.	8—22	1	1	4	3	7	6	6	2	5	6	—	6	34
Haddock (large)	15—	—	2	5	10	15	3	12	—	2	5	—	5	39
Do.	10—14	—	3	22	30	52	27	32	27	2	20	6	27	170
Do. (small saleable)	8—9	—	—	19	12	31	8	6	—	18	—	11	11	74
Whiting (large)	14—	—	6	3	5	8	1	3	—	—	10	—	10	22
Do.	12—15	—	39	24	23	47	15	31	16	6	22	2	24	219
Do. (small saleable)	7—9	—	—	10	11	21	13	53	13	3	—	—	—	102
Hake	12—	—	15	44	106	150	73	56	56	48	26	2	28	404
Ling	15—	—	—	1	1	2	2	2	2	1	2	1	3	12
Lythe	15—	—	—	—	6	6	—	2	—	—	—	—	—	8
Saithe	8—	—	—	—	1	1	—	2	59	—	—	—	—	62
Sail-Fluke	7—	—	30	11	3	14	25	1	5	9	8	—	8	92
Witch	7—	2	176	522	713	1235	650	257	537	338	309	1	310	3285
Long-Rough Dab (large)	10—	—	59	2	3	5	2	—	6	—	2	—	2	74
Do.	8—9	—	4	37	67	104	35	74	25	32	79	—	79	373
Turbot	7—	—	2	1	1	2	1	1	—	2	3	—	3	10
Brill	7—	—	2	2	1	3	6	3	—	4	2	—	2	20
Plaice (large)	15—	1	5	2	2	4	—	1	7	2	4	—	4	24
Do.	12—18	13	48	21	32	53	5	37	25	16	15	—	15	212
Do. (small saleable)	7—11	—	—	3	7	10	9	3	4	5	9	1	10	41
Dab (large)	12—	1	6	—	—	—	—	1	—	1	—	1	1	10
Do.	7—11	14	125	65	56	121	53	66	97	48	77	27	104	628
Lemon-Dab	7—	11	57	80	65	145	59	65	38	46	32	36	68	509
Sole	7—	3	8	4	6	10	12	5	7	4	9	—	9	58
Flounder (large)	10—	—	1	—	5	5	—	—	—	—	—	—	—	6
Do.	7—9	4	21	—	—	—	—	—	—	—	—	—	—	25
Gurnard (large)	11—	—	—	40	119	159	25	33	57	127	89	—	89	490
Do.	7—10	—	—	200	150	350	157	93	70	122	187	6	193	985
Red Gurnard	7—	—	—	8	9	17	9	2	1	3	—	—	—	32
Sapphirine Gurnard	7—	—	—	—	—	—	—	—	—	—	8	—	8	8
Bream	12—	—	—	—	—	—	—	1	—	—	—	—	—	1
Conger	13—	—	—	—	3	3	1	1	—	1	2	—	2	8
		50	655	1038	1585	2623	1247	854	910	857	957	95	1052	8418

* Number of Hauls.

III.

RITH OF CLYDE.

FOR FIVE YEARS.

JULY	SEPTEMBER	OCTOBER	NOVEMBER			Totals of each Species
1890	1897	1896	1895	1896	Totals	
12	12	11	12	1	13	97
—	1	—	—	—	—	35
13	41	49	23	2	25	189
—	—	—	1	—	1	1
—	—	11	2	—	2	22
—	—	—	—	—	—	1
—	14	—	—	—	—	14
—	—	—	—	—	—	2
—	—	—	—	—	—	3
1	—	—	—	1	1	16
—	1	—	23	2	25	68
10	1	—	—	—	—	11
19	19	68	7	—	7	293
1	3	—	—	—	—	7
138	106	129	159	5	164	1042
—	—	—	—	—	—	2
99	96	92	92	40	132	789
3	2	4	2	—	2	52
3	24	44	13	6	19	430
—	—	—	—	—	—	5
—	—	—	—	—	—	1
7	56	98	18	1	19	246
1	—	1	—	—	—	7
—	—	—	—	—	—	2
1	—	—	—	—	—	1
—	4	—	—	—	—	4
—	—	—	—	—	—	1
—	1	—	—	—	—	3
—	1	—	—	—	—	1
—	—	—	—	—	—	1
4	14	19	31	1	32	125
13	—	—	—	—	—	39
2	—	1	2	—	2	5
32	7	1	31	—	31	79
—	5	—	—	—	—	5
—	—	1	—	—	—	6
—	—	—	—	—	—	1
347	396	518	404	58	462	3509

TABLE XXVIII.

"GARLAND'S" TRAWLING. FRITH OF CLYDE.

UNSALEABLE FISHES. TOTALS FOR FIVE YEARS.

SPECIES	Size in Inches	FEBRUARY	MARCH	APRIL			MAY	JULY	SEPTEMBER	OCTOBER	NOVEMBER			Totals of each Species
		1888	1888	1896	1897	Totals	1897	1890	1897	1896	1895	1896	Totals	
		2*	14	12	12	24	12	12	12	11	12	1	13	97
Grey Skate	—12	2	29	1	—	1	2	—	1	—	—	—	—	35
Thornback	—13	—	—	12	24	36	25	13	41	49	23	2	25	189
Starry Ray	—11	—	—	—	—	—	—	—	—	—	1	—	1	1
Sandy Ray	—11	—	—	6	—	6	5	—	—	11	2	—	2	24
Flapper Skate	—11	—	1	—	—	—	—	—	—	—	—	—	—	1
Cuckoo-Skate	—11	—	—	—	—	—	—	—	14	—	—	—	—	14
Raia maculata	—11	—	1	—	—	—	1	—	—	—	—	—	—	2
circularis	—11	—	3	—	—	—	—	—	—	—	—	—	—	3
Cod	—8	—	12	1	—	1	1	1	—	—	—	1	1	16
Haddock	—8	—	25	17	—	17	—	—	1	—	23	2	25	68
Whiting	—7	—	—	—	—	—	—	10	1	—	—	—	—	11
Hake	—12	—	107	3	51	54	19	19	19	68	7	—	7	293
Witch	—7	—	—	—	3	3	—	1	3	—	—	—	—	7
Long-Rough Dab	—8	—	1	157	134	291	213	138	106	129	159	5	164	1042
Plaice	—7	1	1	—	—	—	—	—	—	—	—	—	—	2
Dab	—7	—	140	84	63	147	83	99	96	92	92	40	132	789
Lemon-Dab	—7	—	27	6	1	7	7	3	2	4	2	—	2	52
Gurnard	—7	—	265	16	18	34	41	3	24	44	13	6	19	430
Red Gurnard	—7	—	2	—	3	3	—	—	—	—	—	—	—	5
Labrus (?) maculatus	—	—	—	1	—	1	—	—	—	—	—	—	—	1
Poor Cod	—	—	29	14	16	30	7	7	56	98	18	1	19	246
John Dory	—	—	3	—	1	1	1	1	—	1	—	—	—	7
Trigla lucerna	—	—	—	—	2	2	—	—	—	—	—	—	—	2
Smelt	—	—	—	—	—	—	—	1	—	—	—	—	—	1
Argentine	—	—	—	—	—	—	—	—	4	—	—	—	—	4
Ctenolabrus rupestris	—	—	—	—	—	—	1	—	—	—	—	—	—	1
Zeugopterus punctatus	—	—	—	1	1	2	—	—	1	—	—	—	—	3
Liparis montagui	—	—	—	—	—	—	—	—	1	—	—	—	—	1
Cottus scorpius	—	—	1	—	—	—	—	—	—	—	—	—	—	1
Angler	—	—	1	11	14	25	30	4	14	19	31	1	32	125
Dragonet	—	2	9	5	2	7	6	13	—	—	—	—	—	37
Nursehound	—	—	—	—	—	—	—	2	—	1	2	—	2	5
Picked Dogfish	—	—	—	—	5	5	3	32	3	1	31	—	31	75
Spotted Dogfish	—	—	—	—	—	—	—	—	5	—	—	—	—	5
Lesser Dogfish	—	—	—	—	—	—	5	—	—	1	—	—	—	6
Black-mouthed Dogfish	—	—	—	1	—	1	—	—	—	—	—	—	—	1
		5	657	336	338	674	450	347	396	518	404	58	462	3509

* Number of Hauls.

TABLE XXIX.

"GARLAND'S" TRAWLING. FRITH OF CLYDE.

NO. OF HAULS PER MONTH.

Year	Feb.	March	April	May	July	Sept.	Oct.	Nov.	Colder Months	Warmer Months	Total No. of Hauls
1888	2	11	—	—	—	—	—	—	13	—	13
1890	—	—	—	—	12	—	—	—	—	12	12
1895	—	—	—	—	—	—	—	12	12	—	12
1896	—	—	12	—	—	—	11	1	13	11	24
1897	—	—	12	12	—	12	—	—	12	24	36
No. of Hauls per month	2	11	24	12	12	12	11	13	50	47	97
Average No. per Haul	27	119	137	141	100	108	125	116	123	118	121

No. of Hauls	Year	Plaice	Dab	Average per Haul of certain Fishes: Dab	Long-Rough Dab	Gurnard	Hake	Witch	
13	1888	69	286	22	5	20	9	13	February, March
12	1890	41	166	13	17	10	6	19	July
12	1895	28	169	14	20	24	2	25	November
24	1896	50	358	14	15	19	6	35	April, Oct., Nov.
36	1897	91	448	12	17	21	9	47	April, May, Sept.
97	Five years	279	1427	14	15	19	7	34	

TABLE XXX.

"GARLAND'S" TRAWLING. FRITH OF CLYDE.

STATIONS I. TO XII.

No. of Hauls	Year	Totals: Plaice	Dab	Lemon-Dab	Long Rough Dab	Cod	Haddock	Whiting	Gurnard	Hake	Ling	Sail-Fluke	Witch	Turbot	Brill	Flounder	Gray Skate	Thornback	Saithe	Sole	Average per Haul of certain Fishes: Dab	Long Rough Dab	Gurnard	Hake	Witch	
13	1888	69	266	95	64	14	30	81	265	122	—	30	178	1	2	26	40	—	—	11	22	5	20	9	13	February, March
12	1890	41	166	68	112	8	50	96	129	73	2	1	238	1	3	—	1	35	2	5	13	17	10	6	19	July
12	1895	28	169	34	240	6	44	31	269	33	2	8	309	3	2	—	—	51	—	9	14	20	24	2	25	November
24	1896	50	358	172	362	12	100	48	461	145	3	20	861	3	6	—	2	89	—	8	14	15	19	6	35	April, Oct., Nov.
36	1897	91	468	192	611	15	127	97	762	324	5	33	1706	2	7	5	7	169	60	25	12	17	21	9	47	April, May, Sept.
97	Five years	279	1427	561	1489	55	351	354	1905	697	12	92	3292	10	20	31	50	344	62	58	14	15	19	7	34	

TABLE XXXI.

"GARLAND'S" TRAWLING. FRITH OF CLYDE.

AVERAGES AT EACH STATION AND TOTALS.

No. of Hauls	Year	Average per Haul (Saleable and Unsaleable Fishes) at each Station												Totals			Average per Haul at all Stations		
		I	II	III	IV	V	VI	VII	VIII	IX	X	XI	XII	Saleable Fishes	Unsaleable Fishes	Grand Total	Saleable Fishes	Unsaleable Fishes	Saleable and Unsaleable Fishes
13	1888	16	92	17	18	17	89	230	306	179	86	111	114	705	662	1367	54	51	105
12	1890	97	45	148	94	21	258	142	62	139	112	70	13	854	347	1201	71	29	100
12	1895	54	111	18	59	51	78	116	95	108	118	174	379	957	404	1361	79	33	113*
24	1896	123	89	101	79	78	143	46	161	125	229	120	155	1990	912	2902	82	38	120
36	1897	79	75	107	35	32	187	165	205	195	181	170	206	3742	1184	4926	103	32	136
97	Five Years	81	82	88	54	43	159	134	175	157	164	138	179	8248	3509	11757	85	36	121

* Differences due to fractions.

…ALS	NOVEMBER								Totals	
	VII		VIII		XII		Totals			
Night	Day	Night	Day	Night	Day	Night	Day	Night	**Day**	Night
—	53	121	65	64	286	358	404	543	724	1093
548	—	—	—	—	—	—	—	—	1730	1476
548	53	121	65	64	286	358	404	543	2454	2569

TABLE XXXII

"GARLAND'S" TRAWLING. FRITH OF CLYDE.

COMPARISON OF DAY AND NIGHT DATA

Year	April								May								September								November								Total	
	VII*		VIII		XII		Totals		VII		VIII		XII		Totals		VII		VIII		XII		Totals		VII		VIII		XII		Totals			
	Day	Night	Day	Night	Day	Night	Day	Night	Day	Night	Day	Night	Day	Night	Day	Night	Day	Night	Day	Night	Day	Night	Day	Night	Day	Night	Day	Night	Day	Night	Day	Night	Day	Night
[illegible]	[illegible]	[illegible]	[illegible]	[illegible]	[illegible]	[illegible]	[illegible]	[illegible]	[illegible]	[illegible]	[illegible]	[illegible]	[illegible]	[illegible]	[illegible]	[illegible]	[illegible]	[illegible]	[illegible]	[illegible]	[illegible]	[illegible]	[illegible]	[illegible]	[illegible]	[illegible]	[illegible]	[illegible]	[illegible]	[illegible]	[illegible]	[illegible]	[illegible]	[illegible]
Total	[illegible]	[illegible]	[illegible]	[illegible]	[illegible]	[illegible]	[illegible]	[illegible]	[illegible]	[illegible]	[illegible]	[illegible]	[illegible]	[illegible]	[illegible]	[illegible]	[illegible]	[illegible]	[illegible]	[illegible]	[illegible]	[illegible]	[illegible]	[illegible]	[illegible]	[illegible]	[illegible]	[illegible]	[illegible]	[illegible]	[illegible]	[illegible]	[illegible]	[illegible]

* Station.

www.ingramcontent.com/pod-product-compliance
Lightning Source LLC
Chambersburg PA
CBHW032029220426
43664CB00006B/412